Time and Mind

Professor Dr. HEDE HELFRICH
Department of Psychology
University of Regensburg
D-93040 Regensburg
Germany
e-mail: hhelfrich@aol.com
e-mail: s5hehe@rz.uni-jena.de

Time and Mind

edited by
Hede Helfrich

Hogrefe & Huber Publishers
Seattle • Toronto • Göttingen • Bern

Library of Congress Cataloging-in-Publication Data

Time and mind / Hede Helfrich, editor.
p. cm.
Proceedings of the International Symposium on Time and Mind held in Dec. 1994 at the University of Regensburg.
Includes bibliographical references and indexes.
ISBN 0-88937-173-3
1. Time perception--Congresses. 2. Memory--Congresses. 3. Time perception--Cross-cultural studies--Congresses. I. Helfrich, Hede. II. International Symposium on Time and Mind (1994 : University of Regensburg)
BF468.T544 1996
153.7'53--dc20 96-8729
CIP

Canadian Cataloguing in Publication Data

Time and mind

Based on an International Symposium on Time and Mind, held at the University of Regensburg, Dec. 1994.
Includes bibliographical references and index.
ISBN 0-88937-173-3

1. Time - Psychological aspects - Congresses. I. Helfrich, Hede. II. International Symposium on Time and Mind (1994 : University of Regensburg).

BF468.T55 1996 153.7'53 C96-930885-X

USA: P.O. Box 2487, Kirkland, WA 98083-2487
Phone (206) 820-1500, Fax (206) 823-8324
CANADA: 12 Bruce Park Avenue, Toronto, Ontario M4P 2S3
Phone (416) 482-6339
GERMANY: Rohnsweg 25, D-37085 Göttingen
Phone +49 551 496090, Fax +49 551 4960988
SWITZERLAND: Länggass-Strasse 76, CH-3000 Bern 9
Phone +41 31 300-4500, Fax +41 31 300 4590

Printed in the United States of America

ISBN 0-88937-173-3
Hogrefe & Huber Publishers
Seattle • Toronto • Göttingen • Bern

Preface

The study of time has been approached from a variety of disciplines: it has always been a matter of interest in philosophical thinking, and has been a crucial concept in physics. More significant, however, for most people is that time shapes our behavior and our thinking in almost all areas of our daily lives. Biological rhythms have relevance to practical human situations such as shift work and jet lag; social life submits to temporal rules in work and leisure; ritual follows rules of duration, simultaneity and succession; speech, music, and motor skills show fine temporal tuning patterns; and our memory is organized along temporal structures. Besides these largely automated temporal regulations we have conscious time experiences such as anticipation of future events, feelings of time pressure, and retrospective consideration of elapsed time. Thus, time is a phenomenon which encompasses almost all aspects of our daily lives.

For the psychology of time, the publication of Paul Fraisse's book "Psychologie du temps" in 1957 was a milestone. Since then, the phenomenon of time has drawn an increasing amount of attention from psychologists. Moreover, while earlier works focused primarily on the perception, estimation, and recall of empty or filled time intervals, the scope has gradually expanded to a multiperspective view, stretching from basic psychobiological processes to cross-cultural aspects. However, there seems to be a striking dichotomy between, on the one hand, views based upon biological time constraints hardly accessible to the conscious mind, and on the other hand, views based upon the socially and culturally shaped time representations. What seems both absent and needed is a unified conceptual framework capable of integrating the diverse aspects of the subject of time and bridging the gap between biological and cognitive approaches to temporality. This desideratum led to the idea of bringing together researchers from different scientific backgrounds for an International Symposium on Time and Mind where they could present their perspective and discuss steps toward a unified model of psychological time. Fortunately, the Hans-Vielberth-Stiftung of the University of Regensburg supported this idea and provided me a grant to organize the symposium.

The symposium took place at the University of Regensburg in December 1994. It consisted of review lectures, group discussions, and working sessions focusing on four main topics: ***psycho-biological time constraints***, *time memory and perception*, *cultural differences in psychological time*, and *models of psychological time*. The lectures, discussions, and sessions found a strong resonance not only with the invited participants, but also with guests and students of psychology as well as other disciplines. Thus, it seemed worthwhile to make available the issues and ideas discussed during the symposium in the form of a book to a group wider than a few specialists including colleagues from allied disciplines, new researchers, graduate students and advanced undergraduates.

The purpose of the book is to provide a status report on work regarding psychological time. The volume is intended to highlight theoretical positions, developments in methods, critical findings and questions, and to outline promising approaches toward a synthesis of divergent positions.

The book starts with an introductory chapter by MARC RICHELLE, shedding light on the different facets of the psychological study of time and discussing the question of a continuity or rupture between the different levels of complexity ranging from elementary biological timing to conscious representation and conceptualizing of time.

Part II deals with psycho-biological time constraints, or more precisely, with the timing of motor behavior which is usually hardly accessible to the conscious mind. ANDRAS SEMJEN points out that the regularity of temporal patterns in movements does not necessarily imply the existence of a temporal control. Distinguishing between events (such as spoken speech or produced music) referring to intended outputs and movements referring to goal-related processes, he demonstrates, that only the former are pre-programmed, whereas the latter reflect self-organizing processes in the motor system.

A distinction between intended events and goal-related movements is discussed by ALF C. ZIMMER. He addresses the question of how the expectations about motor actions to be performed affect the allocation of time. From theoretical analyses and empirical results he concludes that the timing of motor actions cannot be accounted for by an internal clock, but rather by interaction of perceived actual task requirements, with motor action units shaped by learning processes.

Part III of the book, dealing with memory and the perception of time, contains two divergent views of psychological time: the psychophysical perspective, where psychological time is considered as a systematically altered perception of physical time, and an information-processing approach, where the processing of temporal information is considered as a largely deliberate cognitive activity, requiring much attention and strategic flexibility on the part of the human individual.

HANNES EISLER looks at the perception and memory of time from a psychophysical perspective. Based on a power function of psychophysical judgment which is derived from experiments using the reproduction method, he proposes a parallel-clock model. Instead of any kind of memory, two sensory registers (i.e. clocks) are assumed, one accumulating the perceived duration from the beginning of the standard, the other from the beginning of the reproduction. The two durations are continuously compared until the correct ratio is reached, after which the response (i.e. the button press) is carried out. Eisler discusses the applicability of the model to both humans and lower order biological species.

In contrast to a psychophysical perspective, JOHN MICHON states that psychological experience of time is not a distorted or altered perception of a determinate thing called 'time', but rather encompasses multiple representations constructed by individuals to cope with multifold and changing life situations in the form of temporal planning of actions. According to Michon, formalized theories within the framework of artificial intelligence may provide essential insights into the way humans process temporal information and stay tuned to the dynamic world around them via the planning and executing of actions. As an illustration of this approach the Generic Intelligent Driver Support (GIDS) model, which is specifically designed as a real-time system to support planning, maneuvering, and control in car-driving tasks, is presented.

Part IV of the book looks for cultural differences in psychological time. In a chapter, written by myself, four dimensions accounting for cultural variability are identified: *objective time* referring to the mental representations of the flow and structure of time, *subjective time* referring to the individual time horizon or temporal perspective, *social time* referring to the organization of time in social life, and *biological time* referring both to irreversible biological events and to recurrent biological cycles or rhythms. The chapter ends with an illustration of the possible interaction between cultural factors and the perception of short-time intervals.

The dimension of social time is further investigated by ROBERT V. LEVINE. As the most significant defining element of social time, he identifies the social tempo or *pace of life*, which he describes as the "heartbeat of a culture." Levine has empirically studied the pace of life in terms of walking speed, working speed, and accuracy of clocks in six countries (England, Italy, Indonesia, Japan, Taiwan and the USA) and found strong and consistent differences in all three measures among the individual countries: the highest social tempo was found in Japan, the lowest in Indonesia. Interestingly, a correlation of social speed with coronary heart disease turned out to be culture-specific: a positive relationship was found in the USA, but not in Japan. To account for this result, Levine posits that high speed is associated with competitive hostility in the USA, as opposed to Japan, where competitive hostility plays little part in the hardworking and fast pace.

Part V is characterized by an attempt to bring together the different models of time as an important step on the way towards integration. Two different integration models are proposed: the scalar model and the attentional gate model. An important difference between the two models is the treatment of non-timing factors such as attention or motivation. Lejeune and Richelle tend to treat these non-timing factors as response bias and behavioral noise to be eliminated from "pure time", whereas Block and Zakay consider them as intrinsic components of psychological time.

HELGA LEJEUNE and MARC RICHELLE suggest a specific clock model - the scalar model - as a robust candidate for a general model, especially in unifying human and animal time. They argue that in spite of cross-species differences in the sensitivity of time, the mechanisms of the internal clock could be the same, with the differences being due to differences in the temporal resolution of the clock. Moreover, the scalar model seems to be open to new adjustments such as connectionist networks. However, Lejeune and Richelle concede that both the advantage and shortcoming of the scalar model is that it is a model of "pure time", excluding up to now all non-timing factors such as attentional strategies, motivational processes and linguistic transformations of tasks.

RICHARD A. BLOCK and DAN ZAKAY categorize the existing models of psychological time. They distinguish two classes of currently mutually incompatible models: one proposing timing *with a timer*, and the other timing *without a timer*. Typical for the first group are the clock models, and for the second group, the cognitive models. To reconcile the two approaches, Block and Zakay propose the *attentional-gate model*, which seems to be capable of combining a psycho-biological pacemaking mechanism with cognitive and motivational processes within a single model.

The book has benefitted from the fact that the contributors read and reviewed each other's chapter. I would like to thank them for this collaborative effort. I also thank Katrin Selinger who supported me in preparing the subject index. Special thanks go Erich Hölter, from whom I received indispensable help and support. He not only encouraged me, but also kept track of the ultimate form of the book.

Regensburg, May 1996 Hede Helfrich

Table of contents

List of contributors

RICHARD A. BLOCK, Prof. Dr., Department of Psychology, Montana State University, Bozeman, MT 59717-0344, USA (e-mail: upyra@gemini.oscs.montana.edu)

HANNES EISLER, Prof. Dr., Department of Psychology, Stockholm University, S-106 91 Stockholm, Sweden (e-mail: he@psychology.su.se)

HEDE HELFRICH, Prof. Dr., Department of Psychology, University of Regensburg, D-93040 Regensburg, Germany (e-mail: hhelfrich@aol.com); (e-mail: s5hehe@rz.uni-jena.de)

HELGA LEJEUNE, Dr., Unité de Psychobiologie des Processus Temporels (U.P.P.T.), Université de Liège, 5 bd du Rectorat (B32), Sart Tilman, B-4000 Liège, Belgium (e-mail: helejeune@vm1.ulg.ac.be)

ROBERT V. LEVINE, Prof. Dr., Department of Psychology, California State University, Fresno, CA 93740-0011, USA (e-mail: robertle@zimmer.csufresno.edu)

JOHN A. MICHON, Prof. Dr., Netherlands Institute for the Study of Criminality and Law Enforcement (NISCALE), Leiden, Witte Singel 103, NL-2313 AA Leiden, The Netherlands (e-mail: NSCR@niscale.leidenuniv.nl)

MARC RICHELLE, Prof. Dr., Unité de Psychobiologie des Processus Temporels (U.P.P.T.), Université de Liège, 5 bd du Rectorat (B32), Sart Tilman, B-4000 Liège, Belgium (e-mail: richelle@vm1.ulg.ac.be)

ANDRAS SEMJEN, Dr., Centre de Recherche en Neurosciences Cognitives C.R.N.C, CNRS-CRNC - 31 Chemin Joseph Aiguier, F-13402 Marseille Cedex 20, France (e-mail: Semjen@lnf.cnrs-mrs.fr)

DAN ZAKAY, Prof. Dr., Department of Psychology, Tel Aviv University, Tel Aviv, Israel (e-mail: dzakay@CCSG.TAU.AC.IL)

ALF C. ZIMMER, Prof. Dr., Department of Psychology, University of Regensburg, D-93040 Regensburg, Germany (e-mail: alf.zimmer@psychologie.uni-regensburg.de)

Part I:
Introduction

Chapter 1: The expanding scope of the psychology of time*

Marc Richelle

Time in life: the slow way toward recognition of chronobiology

Time has been a matter of interest since the most remote past of philosophical thinking, and has always been a crucial concept in physics, where it has gone through the most puzzling transformations, eventually reaching, in contemporary physics and cosmology, breathtaking definitions alien to anything we can grasp on the basis of our daily experience. In the living world, time is obviously a pervasive dimension, with two aspects: (1) irreversible progress of living systems - individual organisms, species - toward degradation and death, while mechanisms of reproduction and speciation perpetuate life; and, (2) as long as the systems keep functioning, cyclic repetitive mechanisms. The first aspect has been central in all fields of biology dealing with evolution, development, and the genetic bases of growth and aging processes. The second aspect has been the theme of chronobiology. Now a recognized subfield of biology, chronobiology, or the study of periodicities and rhythms in living systems, had been for many years the esoteric domain of a handful of pioneers, until the fifties when it was admitted in academic biology circles. It took more years until it eventually found the place it deserves in the general information of biologists and physicians, and in the scientific culture of laypeople.

As often occurs, practical issues were determinant in promoting that particular approach to biological phenomena long considered as at

---

* The present paper has benefited by the support of the Belgian Ministry of Scientific Policy, under project number 31, „Pôle d'Attraction Interuniversitaire“, Temporal Reasoning and Behavioral Variability, granted to the University of Liège.

the fringe of biological science. For example, the physiological consequences of shift-work and of jet-lag, important variations in the toxicity of some therapeutic compounds as a function of time of administration in the circadian cycle are typical cases which drew attention to biological phenomena the importance of which had been proclaimed long before by scientists pioneering in chronobiology. Biological cycles are now duly recognized and an impressive amount of research activity, both basic and applied is devoted to them.

This short reminder of the time it took to have time accepted as an important aspect of biology has a double purpose in this opening chapter. First, we shall have to relate the psychological study of time to chronobiological knowledge. The subjects studied by psychologists, be they animals or humans, are living organisms, and when dealing with their capacities to time their actions or to process temporal information, we cannot ignore the question of the relationship of whatever we observe at the behavioral or cognitive levels with the timing mechanisms studied by chronobiologists. Second, one cannot help noting the parallel between the slow acceptance of the contributions of chronobiology and the scarcity of psychological research on time until very recently. The analogy is no explanation, of course, but it might give psychologists of time some reasonable hope as to their future.

Psychology of time: a brief retrospective

Let us briefly look at the history of the field. To do this, we must first clarify what the psychology of time is about, at least in general terms - more refined comments on that will come later on. Obviously, time has not been ignored by psychology if we consider it as an independent variable. Since the origin of scientific psychology, a lot of things have been studied as a function of time, be it time elapsed since presentation of the stimulus in a learning or memory task, time of presentation of the stimulus in an identification task using the tachystoscope, time in months or years throughout development, etc. Time psychology, to put it in a somewhat simplified way, deals with time as a dependent variable, or in other words it studies the subjects' capacities in keeping time, in estimating time, in regulating their own actions in time, in perceiving and conceiving time. What looks closest to time psychology in early experimental psychology is of course the reaction time. But in most cases it was used for other purposes, and its most popular application in modern psychology, mental chronometry, consists in taking a measure of time as reflecting something else, in that case the complexity of some (serial) information processing.

For about one century, roughly from 1850 to 1950, time has been the occasional concern of a handful of psychologists, experimenters such as Vierordt (1868), a pioneer in the psychophysics of time, or Pavlov, who showed that animals are able to estimate and discriminate duration; psychopathologists such as Janet (1928), developmentalists such as Piaget (1946) or philosophers-psychologists like Guyau (1890), not to mention Bergson (1896; 1922). Practically no psychologist in those days devoted the major part of his scientific activity to the study of time. The first expert on time is unquestionably Fraisse. His experimental and theoretical work developed from the late forties gained wide recognition, being the most referred to in all the studies on time which followed.

Whatever the possible explanations for the neglect of time as an area of study among psychologists (for some speculative interpretations, see Richelle, 1979; 1993), the factors which favored the increasing interest in the topic from the sixties on are fairly easy to identify. The development of chronobiology was certainly one of them. Chronobiologists were by tradition people open to pluridisciplinary approaches, and having managed to have their science gradually recognized by biology at large, they expressed curiosity for other aspects of time in living systems and were instrumental in having psychologists working on time meet together (for an early example of such gathering, see Ajuriaguerra, 1968).

Another factor was technical. The study of psychological time, at least in some of its basic aspects - psychophysics of time in humans and animals, fine timing of motor behavior - requires refined measuring equipment and efficient methods for data analysis. Early chronometric devices would not do. The expansion of research on psychological time was linked with the development of electronic equipment and computer technology. These were the conditions for the fast growing field of animal studies (Richelle & Lejeune, 1980), which in turn contributed to renewed interest in human studies.

A third factor, although less obvious, might have been the receptivity of cognitive psychology to central issues in the psychology of time. The study of time estimation in animals and humans had led to assume the existence of some internal clock, underlying observed behavioral adjustments to time, in the same way as chronobiologists had hypothesized a biological clock to account for biological rhythms. The search for the neural substrate of the internal clock not having produced yet any conclusive evidence, the stage was free for functional models of time processing similar to the information processing models developed in cognitive psychology. Important parts of the material presented and discussed in the present volume pertain to that vein.

For the last twenty five years or so, the field of the psychology of time has been exploding, as evidenced by exponential jumps in publications and scientific meetings. Two major features of this expansion should be pointed out.

First, a number of new avenues of research have been opened, for example, animal psychophysics of time, the study of fine timing in motor skills or music, temporal organization in memory processes. Secondly, the field is typically pluridisciplinary: not only students of all various aspects of psychological time seem eager to meet and share their problems, and they integrate and cross-fertilize their data and theoretical ideas (major sources now available are characteristically along those lines: see Gibbon & Allan, 1984; Michon & Jackson, 1985; Block, 1990a; Friedman, 1990; Macar et al., 1992). The present volume is in the same spirit.

Psychology of time today

An exhaustive enumeration of all facets of time psychology today would take more space than available. A tentative categorization of the major areas of research, based on criteria related with the specific object of study and/or the particular methodological approach used, will undoubtedly be more informative and a better introduction to the questions which typically arise from confronting the different areas. The chart in Figure 1 (p. 7) and 2 (p. 8) will help representing the various categories and the connections between them. The categorization is overtly hierarchical, in the sense that it is based on the assumption that there are successive levels of complexity from the biological to the cognitive and eventually the socio-cultural aspects. The theoretical issues behind this assumption will be briefly discussed in the conclusions (especially the problem of continuity or discontinuity from one level to the next and the problem of the possibly intricate cross-relations between a given level and other super- or subordinate levels).

Psychological chronobiology

Closest to chronobiology, or more correctly phrased, part of it, the study of the behavioral expression of biological rhythmicity has been explored since the pioneering work of Curt Richter (1965) and complemented by the study of the effects of rhythms' perturbation on psychological functions (Leconte & Lambert, 1990). The latter has obvious relevance to practical human situations such as shift work, jet lags, and prolonged isolation with few or unusual temporal cues (Conroy & Mills, 1970; Colquhoun, 1971).

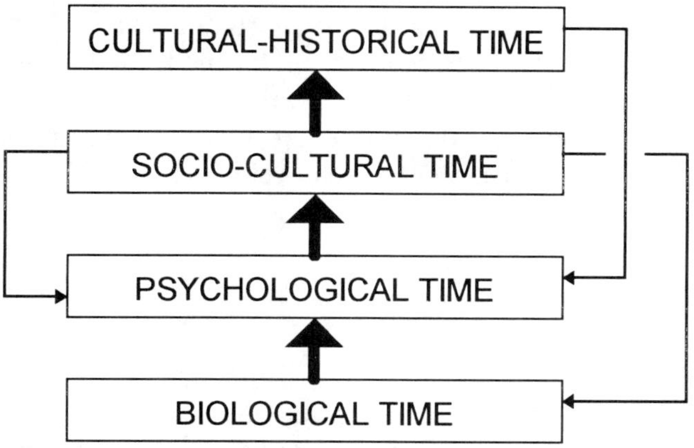

Figure 1: A schematic diagram of levels of temporality in humans, from biological to cultural-historical time. Arrows indicate plausible influences, including interactive loops between levels. Central vertical arrows reflect the assumption that each level is, in some way, an emergent product of the lower levels. „Feedback" arrows indicate the action of higher levels on lower levels (for example: distortion of biological rhythms by mythical or historical constructs). The clear-cut distinction between boxes, especially *psychological time* and *socio-cultural time*, is for didactic purposes.

Temporal regulations of behavior

Behavioral adjustments to durations departing from biological periodicities have been systematically studied since the fifties taking advantage of the technical facilities offered by operant conditioning procedures. These apply equally well to the study of timing behavior - i.e. the distribution by a subject of its own motor responses under specific temporal constraints or contingencies - and to the study of time estimation of external stimuli - an efficient approach to animal psychophysics, merging with human psychophysical studies of time referred to in the next paragraph. We group the whole set of such behavioral adjustments under the generic label *temporal regulations of behavior* (Richelle, 1962). In a first stage, this area has produced essentially descriptive data (Richelle & Lejeune, 1980). More recently models have been developed to account for the internal operations plausibly at work when a subject exhibits temporal regulations. One of the most popular of such models, called the *scalar model*, has been elaborated by Gibbon and Church (Gibbon, 1977; Gibbon & Church, 1981). It has undergone several revisions, the last one bringing a rather drastic change from a cognitivist to a connectionist model (Church & Broadbent, 1990a and b). It has also been used as a reference in attempts at integrating data and concepts from other areas, especially the cognitive processing of time in human memory and reasoning. Chapter 8 in the present volume exemplifies such an endeavor.

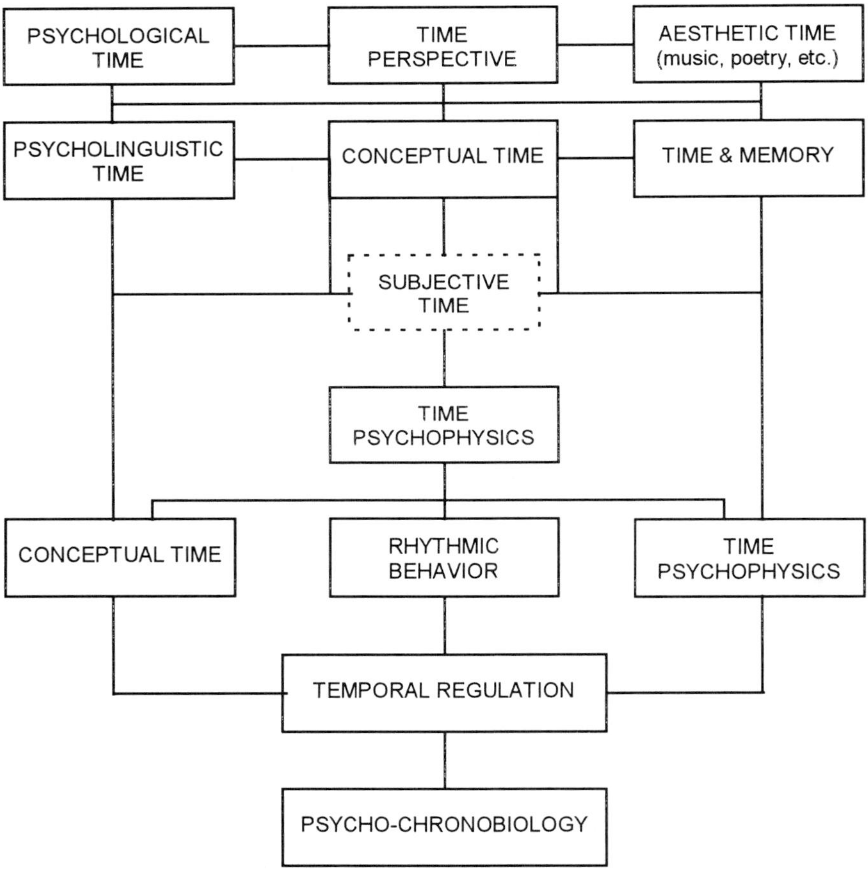

Figure 2: A simplified representation of the main subfields in the psychology of time. A rich network of connections is assumed between boxes. Subjective time is taken as the hypothetical limit between what is common to humans and animals and what is specific to humans.

Psychophysics of time

Early studies on time estimation have been carried out in the framework of psychophysical measurement and exclusively on humans. That tradition of research developed rapidly as better equipment for control and measurement of temporal parameters became available. The current stage of development in that area is illustrated in chapter 4, where the sophisticated methods in use and the rich array of empirical data are described.

Both psychophysics and the study of temporal regulations provide tools for a comparative, i.e. cross-species approach to psychological time. Although it has not been developed as one might have expected, given the technical facilities now available, this line of research has received attention from a few psychologists and ethologists (Richelle & Lejeune, 1980; Lejeune & Wearden, 1991; Wearden & Lejeune, 1993). Some of the data and of the questions they raise are presented in chapter 8.

Timing of motor action

Although the emphasis in cognitive psychology has been put on information, to the expenses of motor action, timing is also an important aspect of movement, especially skilled motor behavior such as catching a tennis ball, or writing, and it has received due attention from experts in motor skills (Thomassen et al., 1992). Because of the intrinsic dynamic properties of any motor act, so-called neural network models as developed in the framework of connectionism have appeared especially attractive to students of timing in movements (Grossberg & Kuperstein, 1989; Bullock & Grossberg, 1991; Gaudiano & Grossberg, 1992). In its most refined form, motor timing can take expressive value as in musical performance, an area in which pioneering and promising studies have been carried out in the recent years (Gabrielsson, 1987; Shaffer, 1989; see also Semjen and Zimmer, this volume, chapters 2 and 3).

Subjective time and the idea of time

From time estimation and timing behavior captured at the basic level typical of the above described areas, psychology of time goes one step further to what would appear exclusively human features, the idea of time and the experience of time. The latter might appear as the subjective side of sheer estimation of duration, while the former might be thought of as a more elaborate product of reasoning activity. These two levels have been shown to be more difficult to disentangle from one another than it would first appear.

The subjective experience of time is itself a puzzling issue, with a number of unresolved contradictions. For example, people may experience a rather short time period as long if it is filled with a lot of unpleasant events or actions, or they may experience a rather long time interval as short if it is filled with a number of pleasant events.

Quite obviously, what we call our experience of time is rarely a spontaneous, intuitive estimation of time elapsed. More often than not, it is contaminated by elaborate inferences of a cognitive nature, not necessarily conscious.

Cervantes' hero, Don Qixote, reported to his faithful servant Sancho Pança, all the marvelous things he had seen in the magical cave of Montesinos, and the interesting conversations he had with the Lord of the place, estimating that he had been there for three days and three nights. Upon a question by Sancho, he confessed that he had not eaten any meal nor felt the need for it. Knowing how difficult it was to persuade his master - whom he knew had just taken a nap for a couple of hours - Sancho pretended to admit that eating was not needed in the enchanted world Quixote had visited. Subjective experience of time can be influenced by awareness of having missed some periodic physiological constraint (Cervantes, 1615).[1]

In another case, not literary but real, Michel Siffre, the French speleologist and scientist, having stayed underground in a cave for more than one hundred days, estimated his stay to be approximately half of that. Having no time keeping device and no synchronizer, he had inferred his calendar from the succession of sleeping periods and meals. However, he was not aware that, while he had slept with remarkable circadian regularity, he had spaced his main meals to the point of taking only one instead of two per (circadian) day. Consequently, he had interpreted one out of two successive full eight hours sleeping periods as an after lunch nap and the second one as a night (i.e. after dinner) sleep (Fraisse et al., 1968).

In both cases, subjective duration was influenced by cognitive factors, which, in both cases, were responsible for distortion relative to „objective" time.

The subjective experience of time or to phrase it slightly differently, the conscious estimation of duration in humans, is not a primitive intuition of the passage of time, but on the contrary it is the result of an

---

[1] The reference to Don Quixote was provided to the author by Pr Marino Perez, of the University of Oviedo, who should be credited for its use as an example here.

elaborate inference process, which takes into account a number of variables such as number and quality of events occurring during the estimated time period, and the complex relationship between other aspects of things going on. Subjective experiencing of time duration is still central in the current researches in cognitive psychology of time, especially on the temporal organization in memory.

The view that the human experience of duration is a complex, elaborate cognitive product has been strongly argued by Michon (1990), whose current thinking is explored in chapter 5.

Early work by Piaget (1946) emphasized the complex relation between speed, distance and duration operations, which the child does not master until it reaches the Concrete Operations Stage of development. Using the same approach, but exploring other experimental situations, post-Piagetian studies provided a picture even more complex of the evolution of the concept of time with age (Montangero, 1977; 1984; 1993; Levin, 1977; 1992; Crépault, 1989).

Developmental studies in the last ten years have not been limited to the post-Piagetian frame-work: modern cognitive theory provided the background for important developmental contributions (Friedman, 1982; Pouthas, 1985), as well as the behavioral tradition in a comparative perspective (Lejeune, 1993b; Lejeune & Richelle, in press).

Models of psychological time

Theory building has been for some time re-labeled modelization. Models of psychological time have been proposed, as already mentioned, in the frameworks of psychophysics (Allan, 1992; Allan & Kristofferson, 1974) and temporal regulations (Killeen & Fetterman, 1988); the two eventually merging into more general models (Gibbon & Church, 1981; Church & Broadbent, 1990 a and b).

Cognitive models of psychological time - under both aspects of duration and order - such as the one proposed by Block (1990b) are essentially based on information processing and memory. Recent revisions of that model are discussed in chapter 9 of the present volume. Interestingly, current efforts aim at integrating models borrowed from animal research or psychophysics of time and models elaborated in the context of cognitive - information processing type - human psychology (Wearden & Lejeune, 1993; Wearden, 1993). Moreover, as can be expected from the *Zeitgeist*, connectionist approaches are increasingly in favor. Such models, on the one hand, provide for a rapprochement with neurobiology (Miall, 1989; 1992; 1993; Ivry & Keele, 1989). On the other hand, paralleling a general trend in cognitive psychology, they shift the emphasis from a static concept of representation of time toward a dynamic property which links the motor interaction of subjects and their environment, making for an integration with the study of motor timing mentioned above (Freyd, 1992; see Semjen, this volume, chapter 2).

Time, language and culture

Human experience of time cannot be conceived, of course, without considering the role of language. The progressive mastery of temporal linguistic markers was studied by developmental psycholinguists (Bronckart, 1976; Ferreiro, 1971). Recent studies on time reasoning in adults usually involve essentially verbal reasoning (De Vooght & Vandierendonck, 1993). Linguistic tools are a necessary condition for temporal organization in retrospective and prospective memory. Keeping track of order and duration of past events, as well as anticipating and planning one's future is inconceivable without language. However, the way an individual relates his past and present experiences, and anticipates the future are encompassed under the label *time perspective* or *time horizon* (in French: *horizon temporel*). This cannot be reduced to reasoning, be it verbal or not; it is an integral aspect of personality and has been appropriately studied as such (Nuttin, 1979; see also Helfrich, this volume, chapter 6). Personality is largely dependent upon social and cultural factors. Therefore, individual time perspective should be approached in its relation to what could be called collective time perspectives in a given society or culture. Intercultural differences with respect to time have long been observed by sociologists and anthropologists (Durkheim, 1912; Hubert, 1905; Whorf, 1956). The renewed and systematic interest in time as a subject in various scientific fields, has led to abundant production (see Gasparini, 1990) again typically characterized by pluridisciplinary orientation. Research has been focused either on the comparison of time perspective and attitude toward time (e.g., punctuality, short versus long term planning) in different cultures or on the various facets of social time in a given society, usually Western culture. Chapter 7 provides an illustration of the former approach. The latter deals with issues such as time allocation of human activities (work, leisure, social versus individual activities, etc.) in different social groups or classes, and at different life stages; time span of anticipatory behavior (saving and retirement plans; health security planning; consumers' behavior in buying goods or housing, etc.). Quite interestingly, this field of studies can be expected to integrate more and more psychological, individual dimensions on the one hand, and economic and political aspects on the other. Many questions arise from such studies which have implications far beyond science proper (see also Helfrich, this volume, chapter 6). For example, in modern democratic states, blatant contradictions are observed between the recognized need for long term planning with respect to pollution, demography, health security, and the like, and the actual short term management of political life, with a time perspective to the next election usually not extending to more than four or five years; or between the encouragement given to citizens to foresee and prepare their future and the future of their children and the commercial pressure to consume goods before having the resources to pay them. Such contradictions remind one of the Freudian conflict between the pleasure and reality principles. They are contemporary versions of the old problem of (in)compatibility between individual trends, interests, instincts, or motivations and social controls. The current situation may be characterized by subtle arrangements which mask the strength of the controls and create an illusion of freedom. However, whatever social relations of power exist, asymmetrical or reciprocal, time management is a crucial aspect.

Looking ahead

The above survey provides a sample of the numerous facets offered by the study of psychological time today. It suffices to convey an adequate picture of the wide variety of topics dealt with, which range from the sense of time in laboratory rats to the expressive value of timing notes in performing instrumental music, from the neural substrate of time estimation to the time perspectives in modern industrialized societies. The variety of subjects submitted to empirical inquiry ranges from animals of various species and ages, to humans, ranging from new-borns to elderly people. Studies involves a wide variety of cultures; and the diverse methodological approaches are put to work in exploring all possible aspects of time - psychophysical, developmental, comparative, cross-cultural, computer simulation, etc. Such diversity will eventually converge into one integrated theory. Confronting data and concepts from those multiple sources is a condition toward such a theoretical endeavor. A general theory of psychological time will have to answer a number of questions, some of which have been formulated for some time already and some of which will present themselves as research progresses. Of course, only the former can be stated now. We shall not give a complete list of them. We shall limit ourselves to a few, selected questions which require a pluridisciplinary approach and the multiple methods of attack which has been emphasized, and implicitly advocated in the present introductory chapter.

A first, crucial question bears on the continuity-discontinuity issue. That psychological time exhibits different characteristics at different levels of complexity is beyond question. But does it follow that these levels should be viewed as distinct and fully independent from each other or should we look for continuities in order to explain at least part of their genuine properties? In other words, are biological rhythms one key (if not the key) to acquired temporal regulations in animals, and to time estimation and timing in humans, and in turn are these basic capacities the condition for more elaborate, cognitive processing of order and duration? Or, to the contrary, can we better understand higher levels of psychological time perceptions if we focus on their independence with respect to lower levels. For example, the processing of duration by humans can be conceived of as directly derived from what it is in animals, or it can be viewed as an elaborate cognitive construct with little or no resemblance with more elementary processes, whatever its past evolutionary origin (possibly not unrelated to animals' time mechanisms).

The continuity-discontinuity issue has obvious consequences on the place given to the comparative approach, but altogether it cannot be solved without resorting to it.

A second question has to do with the limits and constraints in the psychological processing of time, under its double aspect of duration and order. The issue has very practical implications in humans, and is only beginning to receive the attention it deserves. The problem at stake will be most easily understood by analogy with the neighbor field of chronobiology. It is generally admitted that humans, although they have deeply modified their environment with respect to synchronizers, have not freed themselves from the constraints of biological rhythms. Concern with disturbances of natural periodicities in work or school situations is now quite popular. Similarly, one may ask whether there are limits beyond which the psychological time processor(s) would go astray, and if so what exactly are the constraints. Examples can be taken from very different fields, and with various individual or social implications.

A number of control tasks by human operators (e.g., in industry, air traffic control, space navigation, etc.) involve highly complex time parameters, both in term of sequencing and duration of events. What is the maximal load that can be accepted for safety and for the operator's well-being (De Keyser, 1990)?

Awareness of future risks for the human species motivates long term planning of societies far beyond the present generation; such awareness, however, is currently limited to very few individuals or groups.It is not clear that such awareness can be extended to the population at large and be implemented in changed opinions and acts, which imply management of time of a kind diametrically opposite to what is encouraged by many other societal factors. Are human persons able to behave according to remote goals a few centuries ahead? Are they capable of solving the contradictions in time perspectives to which they are exposed?

Another example of less far reaching consequences can be drawn from the field of music: rhythms and timing have been major component of music throughout ages in most cultures. Musical creation has been exploring new styles, occasionally reducing the traditional salience of rhythmic and temporal cues. To what degree will the average humans be receptive to such musical production?

One might like to phrase the problem at stake here by asking: is there a place for a psychopathology and a sociopathology of time, which would have to locate and describe disturbances differing by far from the loss of time orientation classically described in clinical handbooks?

A third question is about the biological substrate of psychological time. No one would today deny that, however complex, processing of time is implemented in the nervous system. While it is perfectly sound to study psychological time at the behavioral or at the cognitive level, in a functionalist approach (without explicitly referring to the nervous system), it is equally clear that psychological descriptions at some point must correspond with neurobiological studies, which in turn cannot seriously progress without referring to psychological descriptions. In this dialogue, simulated models already play an important part, providing common ground to test new hypotheses. Here again, the current state of affairs makes pluridisciplinarity a necessity.

The same spirit of open pluridisciplinarity is reflected in the content of the present volume. However limited, the sample is a very significant one, in the sense that it includes, besides the cognitive, „information processing" and „representational" approach which has been in favor in the last years, the complementary and no less essential areas are explored, for example, time in movements; individual psychophysical investigations; the consideration of cultural factors; discussion of highly abstract models, and the description of mere facts in animals or humans which sometimes resist to fitting the models.

References

Ajuriaguerra, J. de (1968). *Cycles biologiques et psychiatrie*. Paris: Masson.

Allan, L.G. (1992). The internal clock revisited. In F. Macar, V. Pouthas, & W.J. Friedman (Eds.), *Time, action and cognition: towards bridging the gap* (pp. 191-202). Dordrecht, Netherlands: Kluwer Academic.

Bergson, H. (1896). *Matière et mémoire*. Paris: Alcan.

Bergson, H. (1922). *Durée et simultanéité*. Paris: Alcan.

Block, R.A. (Ed.) (1990a). *Cognitive models of psychological time*. Hillsdale, NJ: Lawrence Erlbaum.

Block, R.A. (1990b). Models of psychological time. In R.A. Block (Ed.) *Cognitive models of psychological time*. Hillsdale, NJ: Lawrence Erlbaum, pp. 1-35.

Bronckart, J.P. (1976). *Genèse et organisation des formes verbales chez l'enfant*. Bruxelles: Dessart et Mardaga.

Bullock, D. & Grossberg, S. (1991). Adaptive neural networks for control of movement trajectories invariant under speed and force rescaling. *Human Movement Science, 10*, 3-53.

Cervantes (1615). Don Quichotte de la Manche [titre original en espagnol à fournir].

Church, R.M. & Broadbent, H.A. (1990a). Alternative representations of time, number and rate. *Cognition, 37*, 55-81.

Church, R.M. & Broadbent, H.A. (1990b). A connectionist model of timing. In M.L. Commons, S. Grossberg & J.E.R. Staddon (Eds.), *Quantitative models of behaviour; neural networks and conditioning* (pp. 225-240). Hillsdale, NJ: Erlbaum.

Colquhoun, W.P. (1971). *Biological rhythms and human performance*. New-York: Academic Press.

Conroy, R.T. & Mills, J.N. (1970). *Human circadian rhythms*. London: Churchill Livingstone, 236 p.

Crépault, J. (1989). *Temps et raisonnement. Développement cognitif de l'enfant à l'adulte*. Lille: Presses Universitaires de Lille.

De Keyser, V. (1990). Temporal Decision Making in Complex Environments. In D.E. Broadbent, J. Reason & A. Baddeley (Eds.). *Human factors in hazardeous situations*. Oxford: Clarendon Press.

De Vooght, G. & Vandierendonck, A. (1993). *On the spatial representation of time concepts in reasoning tasks.* University of Ghent. Reports of the Department of General Psychology. Report 10.

Durkheim, E. (1912). *Les formes élémentaires de la vie religieuse. Le système totémique en Australie*. Paris: Alcan.

Ferreiro, E. (1971). Les relations temporelles dans le langage de l'enfant. Genève: Droz.

Fraisse, P. (1948). Étude comparée de la perception et de l'estimation de la durée chez les enfants et chez les adultes. Enfance, 1, 199-211.

Fraisse, P. (1956). Psychologie du rythme. Paris: Presses Universitaires de France.

Fraisse, P. (1957). *Psychologie du temps*. Paris: Presses Universitaires de France.

Fraisse, P. (1963). *Psychology of time*. New-York: Harper & Row.

Fraisse, P., Siffre, M., Oleron, G. & Zuili, N. (1968). Le rythme veille-sommeil et l'estimation du temps. In *Cycles biologiques et psychiatrie,* Symposium Bel-Air III, Genève, Septembre 1967. Genève: Georg; Paris: Masson.

Freyd, J.J. (1992). Dynamic representations guiding adaptive behavior. In F. Macar, V. Pouthas & W.J. Friedman (Eds.), *Time, action and cognition: towards bridging the gap* (pp. 309-323). Dordrecht: Kluwer Academic Publishers.

Friedman, W.J. (Ed) (1982). *The developmental psychology of time*. New York: Academic Press.

Friedman, W.J. (1990). *About time: inventing the fourth dimension.* Cambridge, Mass.: MIT Press.

Gabrielsson, A. (1987). Once again: the theme from Mozart's piano sonata in A major (K.331): a comparison of five performances. In A. Gabrielsson (Ed.), *Action and perception in rhythm and music* (pp. 81-103). Stockholm: Royal Swedish Academy of Music,.

Gasparini, G. (1990). *Tempo, cultura, societá.* Milano: Franco Angeli.

Gaudiano, P. & Grossberg, S. (1992). Adaptive vector integration to endpoint: Self-organizing neural circuits for control of planned movement trajectories. In A.J.W.M. Thomassen, D.A. Rosenbaum & P.C.W. van Wieringen (Eds.), *Human Movement Science, 11*, 141-155.

Gibbon, J. (1977). Scalar expectancy theory and Weber's law in animal timing. *Psychological Review, 84*, 279-325.

Gibbon, J. & Allan, L.G. (1984) (Eds.). Timing and time perception. New-York: *Annals of the New York Academy of Sciences, 423.*

Gibbon, J. & Church, R.M. (1981). Time left: Linear versus logarithmic subjective time. *Journal of Experimental Psychology: Animal Behavior Processes, 7*, 87-108.

Grossberg, S. & Kuperstein, M. (1989). *Neural dynamics of sensory-motor control: Expanded edition*. Elmsford. NY: Pergamon Press.

Guyau, J. M. (1890). L'idée de temps. In J.A. Michon, V. Pouthas & J. Jackson (Eds.), *Guyau and the idea of time*. Amsterdam: North-Holland.

Hubert, M. (1905). *Etude sommaire de la représentation du temps.* Paris: Imprimerie Nationale.

Ivry, R.I. & Keele, S.W. (1989). Timing functions of the cerebellum. *Cognitive Neuroscience, 1*, 136-152.

Janet, P. (1928). *L'évolution de la mémoire et de la notion de temps*. Paris: A. Chahine.

Killeen, P.R. & Fetterman, J. G. (1988). A behavioral theory of timing. *Psychological Review, 95*, 274-295.

Leconte, P. & Lambert, C. (1990). *La chronopsychologie* (Que Sais-Je? n° 2549). Paris: Presses Universitaires de France.

Lejeune H. (1993a). *Estimation des durées et régulation des conduites*. Thèse de doctorat spécial en psychologie. Université de Liège. Volume 1: *Faits et Modèles*, pp. 188. Volume 2: *Sélection de Publications*, pp. 332. Université de Liège: Pôle d'Attraction Interuniversitaire (Raisonnement temporel et variabilité comportementale).

Lejeune, H. (1993b). Towards a comparative psychology of timing behavior. *Psychologica Belgica, 33*, 229-240.

Lejeune, H. & Richelle, M. (in press). Time estimation: animals and humans compared, *International Journal of Comparative Psychology*.

Lejeune, H. & Wearden, J.H. (1991). The comparative psychology of fixed-interval performance: some quantitative analysis, *Learning and Motivation*, 22, 84-111.

Levin, I. (1977). The development of time concepts in young children: reasoning about duration. *Child Development, 48,* 435-444.

Levin, I. (1992). The development of the concept of time in children: An integrative model. In F. Macar, V. Pouthas, et W.J. Friedman (Eds.), *Time, action and cognition: Towards bridging the gap* (pp. 13-32). Dordrecht: Kluwer Academic Publishers.

Macar, F., Pouthas, V. & Friedman, W.J. (Eds.) (1992). *Time, action and cognition: towards bridging the gap*. Dordrecht: Kluwer Academic Publishers.

Miall, R.C. (1989). The storage of time intervals using oscillating neurons. *Neural Computation, 1*, 359-371.

Miall, R.C. (1992). Oscillators, predictions and time. Dans F. Macar, V. Pouthas, & W.J. Friedman (Eds.), *Time, action and cognition: towards bridging the gap* (pp. 215-227). Dordrecht: Kluwer Academic Publishers.

Miall, R.C. (1993). Neural networks and the representation of time. *Psychologica Belgica, 33 n° 2*, 255-270.

Michon, J.A. (1990). Implicit and explicit representations of time. In R.A. Block (Ed.), *Cognitive models of psychological time* (pp. 37-58). Hillsdale, NJ: Lawrence Erlbaum.

Michon, J.A. & Jackson, J.L. (Eds.) (1985). *Time, mind and behavior*. Berlin: Springer.

Montangero, J. (1977). *La notion de durée chez l'enfant de 5 à 9 ans*. Paris: Presses Universitaires de France.

Montangero, J. (1984). Perspectives actuelles sur la psychogenèse du temps. *L'Année Psychologique, 84*, 433-460.

Montangero, J. (1993). From the study of reasoning on time to the study of understanding things in time, *Psychologica Belgica, 33 n° 2*, 185-196.

Nuttin, J.R. (1979). La perspective temporelle dans le comportement humain. In P. Fraisse, F. Halberg, H. Lejeune, J.A. Michon, J. Montangero, J. Nuttin & M. Richelle, *Du temps biologique au temps psychologique* (pp. 307-364). Paris: Presses Universitaires de France.

Piaget, J. (1946). *Le développement de la notion de temps chez l'enfant*. Paris: Presses Universitaires de France.

Pouthas, V. (1985). Timing behavior in young children: A developmental approach to conditioned spaced responding. In J. A. Michon & J. Jackson (Eds.), *Time, mind and behavior* (pp. 100-109). Berlin: Springer.

Richelle, M. (1962). Action du chlordiazépoxide sur les régulations temporelles dans un comportement conditionné chez le chat, *Archives internationales de Pharmacodynamie et de Therapie*, *140*: 434-439.

Richelle, M. (1993). A concise guide to the temporal maze. In H. Lejeune & F. Macar (Eds.). Temporal Processes (Thematic issue), *Psychologica Belgica, 33,* complete number 2.

Richelle, M. & Lejeune, H. (1979). L'animal et le temps. In P. Fraisse (Ed.), *Du temps biologique au temps psychologique* (pp. 73-128). Paris: Presses universitaires de France.

Richelle, M. & Lejeune, H. (1980). *Time in Animal behavior*. Oxford: Pergamon Press.

Richter, C.P. (1965). *Biological Clocks in Medicine and Psychiatry*. Springfield, Ill.: Charles C. Thomas.

Shaffer, L.H. (1982). Rhythm and timing in skill. *Psychological Review, 89*, 109-122.

Shaffer, L.H. (1989). Cognition et affect dans l'interpénétration musicale. In S. McAdams & I. Deliège (Eds.), *La musique et les sciences cognitives* (pp. 537-550). Bruxelles: Mardaga.

Thomassen, A.J.W.M., Rosenbaum, D.A. & van Wieringen, P.C.W. (1992) (Eds.). *Sequencing and timing of human movement*. Amsterdam: North-Holland.

Vierordt, K. (1868). *Der Zeitsinn nach Versuchen* [Empirical studies of time experience]. Doctoral Dissertation, Universität Tübingen, Germany.

Wearden, J.H. (1993). Decisions and memories in human timing, *Psychologica Belgica, 33 n° 2*, 241-254.

Wearden, J.H. & Lejeune, H. (1993). Across the great divide: animal psychology and timing in humans. *Time and Society*, *2*, 87-108.

Whorf, B.L. (1956). *Language, thought and reality*. Cambridge: M.I.T. Press.

Part II:

Timing in motor behavior

Chapter 2: Emergent versus programmed temporal Properties of movement sequences*

ANDRAS SEMJEN

Introduction

This chapter deals with time and timing aspects of movement production. Since natural behavior is basically sequential, we will consider here movements produced in sequences, rather than single, isolated movements. Within the context of sequential movements, we will focus on human movements, given the overwhelming importance of timing in many human-specific activities.

Time in motor production: a frame for description or a control dimension?

Since every real movement sequence unfolds in both time and space, concepts of time can be used in the first place to *describe* some characteristics of a sequence. For instance, one can record the *order* in which some movement events have occurred, or how *far apart* they were from each other on a time scale. A more sophisticated way of looking at timing consists of searching for *regularities* in the distribution of events over time. There may be *periodicities* in the occurrence of some event(s) or recognizable *patterns* in their relative positions on the time scale. Moreover, periodicities and patterns of timing may show a variable degree of *stability* during repeated performances. Recording temporal properties of this kind amounts to describing events *in time*. This description may however tell us something quite unrelated to the control *of time*. This point can be illustrated by means of the following example.

Let us assume that a person is requested to produce a sequence of regular taps with the index finger on a response key, and to accentuate

* Preparation of this chapter was partly supported by a CNES (Centre National d'Etudes Spatiales) research grant (N°94/CNES/0274).

every fifth tap. Further, let us suppose that the cadence of the taps is to be based on that of a metronome to which the person has to listen before starting to tap. In this experiment, one kind of timing regularity is likely to appear in the length of the inter-tap intervals, which can be expected to be equal or very similar, on the average, to the length of the metronome interval. The mean inter-tap interval would thus depend on the stimulus conditions (metronome frequency) and on the subject's explicit intent to time the taps as instructed. Another kind of timing regularity can be observed in the times during which the finger is in contact with the response key. Note that, in this experiment, the subject does not receive any instruction concerning this aspect of the tapping performance. The contact duration turns out to be systematically longer in the case of the accentuated taps than in the case of the non-accentuated taps. Moreover, the contact time increases as a function of the tapping interval, with both accentuated and non-accentuated taps. (Figure 1 A, p. 25). Examining the contact time in proportion to the inter-tap interval shows, on the other hand, that the relative contact time is inversely correlated with the length of the inter-tap interval. The slope of the function relating the relative contact time to the length of the interval is the same with both the accentuated and non-accentuated taps (Figure 1 B, p. 25). The temporal characteristics of the finger contacts thus show quite systematic variations depending on the tapping rate and accent production. Nevertheless, these regularities need not, and probably do not, reflect the existence of a preselected timing goal. We have no evidence justifying the assumption that finger contact times may be directly controlled timing properties of the finger tapping sequence. They may rather reflect the force production or force regulation dynamics under the (boundary) conditions of a preselected tapping rate.

It therefore seems to be necessary to distinguish between directly controlled or planned timing properties, and indirectly controlled or emergent timing properties, to be able to give a scientific account of movement sequence timing. Timing regularities do not necessarily reveal the existence of a timing program. On the contrary, a timing program presupposes the existence of a timing goal or a set of timing goals. The distinction between directly controlled and indirectly controlled timing properties is not always easy to make, however. It requires distinguishing between two main levels of analysis, namely between movement level analysis and event level analysis.

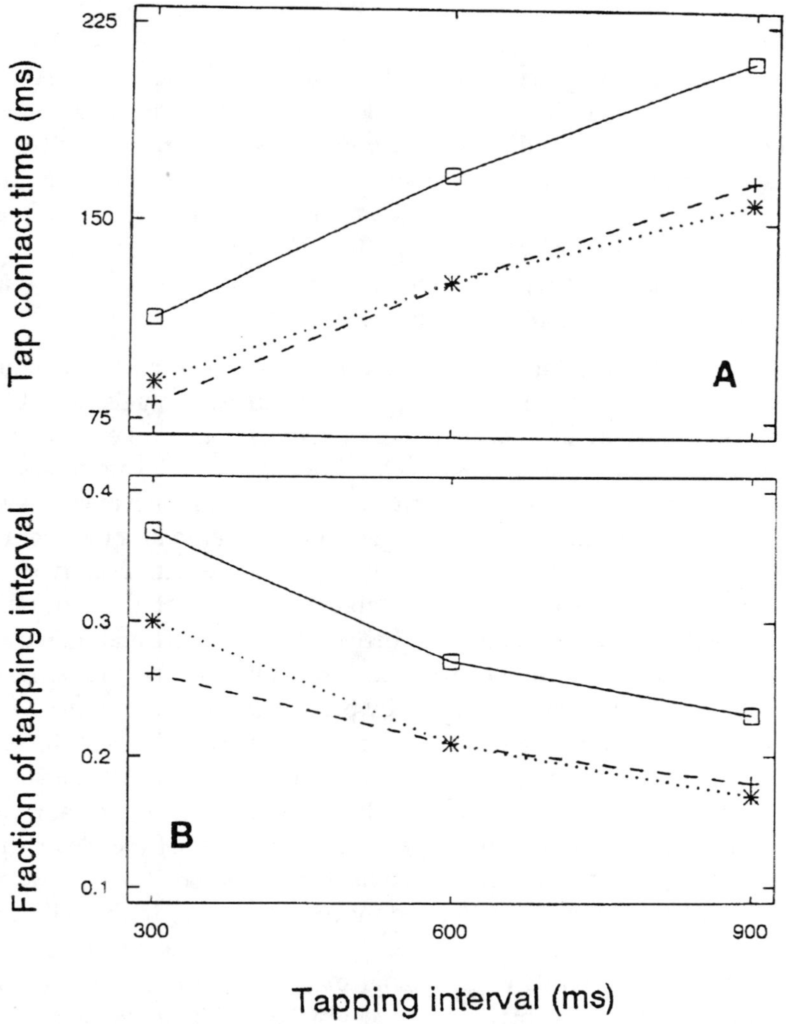

Figure 1: (A): Mean duration (ms) of the contact between the index finger and the response key in regular finger-tapping sequences with intertap intervals of 300, 600, and 900 ms. (B): Mean contact duration expressed as a fraction of the required intertap interval.[1]

[1] Continuous lines: contact duration with forceful taps. Broken lines: contact duration with non-accentuated taps in sequences in which every fifth tap was accentuated. Dotted lines: contact duration in sequences without any voluntary accent.

Levels of analysis

One level of analysis focuses on the variably continuous *displacements* of various body segments, end-effectors, and articulators. At this level of analysis, the timing is embedded in the *kinematics* and *kinetics* of the movement, and in the *coordination* between the space-time trajectories of different movements. Since the movements of our body segments are often oscillatory, timing can be described in terms of *frequency patterns* and/or *patterns of phase relationships*. Patterns of frequency and phase relationships often emerge as the result of self-organizing processes in the motor system (Kelso, 1981).

Another level of analysis deals with the behaviorally meaningful *events* produced by the movements, such as speech sounds, syllables, words or sentences in spoken language production, or keystrokes in typewriting or piano playing. At this level of analysis, relevant timing information may be present in the *inter-event times*, in their means, variability, sequential dependence and organizational hierarchies (Vorberg & Wing, in press). Although behaviorally significant events usually occur concomitantly with some variably stable segment(s) of the movement trajectories, the overall relationship between event timing and movement timing is by no means simple or direct. While the temporal properties of behaviorally important events might by preselected in a *timing program*, that is, in a short-term representation of a set of temporal goals, the relevance of the notion of programming is far more dubious in the domain of movement timing. As we shall see, the event timing can remain very similar despite the fact that the space-time trajectories of the movements involved are prone to variations. In other words, no one-to-one relationship exists between event timing and movement timing. In the following sections, we will consider examples from speech production and the skillful use of keyboards (e.g., typing, piano playing) in order to substantiate the distinction between movement level timing (thought to be the domain *par excellence* of "„emergent“ timing) and event level timing (thought to be the domain of „programmed“ timing).

Speech

Movement level

The production of fluent speech is based on the coordinated activation of a very large number of muscles controlling air volume, air pressure, and the movements of the speech articulators. Coordination is necessary not only between respiration and upper airway activity, but also between the movements of the individual articulators that carry out the compound articulative gestures involved in the production of the acoustic speech signals. A considerable amount of research has been carried out over the past few decades, yielding insights into these coordinative processes (see Abbs & Connor, 1989). For instance, Gracco & Abbs (1986, 1988) have studied the movements of the upper lip, lower lip, and jaw, during the oral closing movements which occur prior to the production of the unvoiced bilabial stop consonant „p" in the word „sapapple." Recordings of the space-time trajectories of the upper lip, lower lip, and jaw showed that the trial-by-trial variability of the movement of each articulator was much greater than the variability of the sum of these movements. A similar picture emerged from the data on the movements' instantaneous velocity. While the combined velocity signal of the three articulators was remarkably stable, the instantaneous velocity of each individual articulator varied considerably from one trial to another. The correlation between the instantaneous velocity of pairs of articulators was also quite weak; whereas the *peak velocity* attained during the movement of each articulator showed a stable temporal pattern of organization. First, there was a fixed ordering of the occurrence of peak velocity regardless of the total movement duration: upper lip (UL), lower lip (LL), and jaw (J). Secondly, the moment of occurrence of the peak velocity was found to be strongly correlated between the articulators, that is, between LL/J, LL/UL, J/UL. Finally, loading the lower lip before and during the oral closing movement did not disorganize the timing pattern of the peak velocities. Most importantly, each individual articulator contributed to some degree to compensating for the load in such a way that the oral closing gesture, which is a prerequisite for the production of the „p" sound, would be properly performed.

In short, both fixed and variable characteristics were found to be present in the production of oral closing before the production of the „p" sound. The acoustic goal was fixed, of course, as was the overall speech gesture necessary to its production. This gesture was performed at variable points in space, as the starting or final position of the articulators, and instantaneous velocity of the latter, varied from trial to trial. The coordination between the movements of the articulators, that is, their *interdependent* generation, was reflected in the fixed ordering of their peak velocities and the proportionate adjustment of their times of occurrence (Gracco & Abbs, 1986).

Precise relative timing of the articulators appears to be an important aspect of their *coordination*. A lack of temporal coordination may cause the speech signal to be altered or completely changed. However, observing that movements are coordinated in time does not entail that time is an independent parameter of movement control. The cooperation between the articulators in the performance of a speech gesture, that is, the coordination between the articulators' movements, is likely to emerge via flexible central couplings between motor output commands, as well as via sensory-based phasic modulation of their interactions (Fowler et al., 1981; Gracco & Abbs, 1988).

Event level

The movements of speech articulators generate variably discrete speech events differing in their size. Speech sounds or *segments* are embedded in *syllables* which constitute *words*; words are assembled into larger units such as *phrases* and *sentences*. Segments, syllables and words can be described in terms of their *absolute or relative duration*. The production of syllables or words may include *periodicities*. Phrases and sentences can be uttered at variable *speaking rate*s. Since speech serves primarily as a means of social communication and is acquired through a process of social interaction, at least some of the above mentioned temporal properties are learned and are therefore typical of the medium common to the given linguistic community. The individual learns what temporal properties may convey semantic or emotional meaning, and how they should be used in an appropriate context. It thus turns out that some of the timing characteristics of speech events are likely to be coded in a timing program because they fulfill a priori timing requirements. One should not however overlook the fact that speech event timing is certainly modulated by motor production constraints (Perkell, 1980).

Speaking rate

The overall rate or tempo of connected speech is subject to considerable variations, both intentional and non-intentional, which depend on extra-linguistic factors such as concentration, familiarity, social environment, and emotional state. The general tempo of speaking may either be pre-set by an internal timekeeper or depend on the overall oscillatory state of the speech motor system. Any variation in speaking rate entails correlated changes in the timing as well as in the phonetic qualities of the speech events at every level, i.e., in the words, syllables and speech segments (sounds). These changes are non-linear, that is, they do not affect all the speech events in the same proportions. For instance, in fast speech the duration of vowels is relatively more compressed than that of consonants (Klatt, 1976); contractions, assimilations and deletions are often constrained by prosodic and syntactic rules (Dogil, 1984). Any *variations* in the actual timing of the speech events as a function of the speaking rate are therefore likely to result from complex interactions between grammatical, prosodic and motor factors.

Besides the overall changes in tempo, the rate of normal speaking may be subject to local variations, i.e., recurrent pauses or declaration in the rhythm of speech production. The *pauses* in normal speech production have been classified on the basis of their *controllability* and the *concern of the interlocutors* (Ballmer, 1978). *Speaker relevant unintentional pauses* have been said to be due for instance to breathing, to the planning of subsequent stretches of speech, to hesitation between contradictory plans, or to searching for words (see also Butterworth, 1980). *Communicative pauses* are by definition intentional and frequently also optional (depending, for instance, on the speaking rate). They may serve to parse the speech stream into its linguistic components, for pauses of this kind have been found to occur at the boundaries of the main syntactic constituents. The length of a pause occurring at the end of a syntactic unit (or alternatively, *the lengthening* of the last syllable in the unit) tends to vary depending on the position of the unit in the syntactic hierarchy: the higher the unit's hierarchical status, the longer the pause, or the greater the lengthening of the terminal syllable will be (Cooper, 1980; Cooper et al., 1978). These variations in the speaking rate can highlight the linguistic structure of the message for the listener and can help, for instance, to disambiguate the meaning of some sentences since they reveal the speaker's actual syntactic codes (Cooper et al., 1978). Pausing can also be used as an expressive tool (stylistic pausing) in order, for instance, to reinforce or emphasize the rhythmic unfolding of speech.

Intentional pausing or the lengthening of the final syllable of a word may be viewed as a kind of „negative timing“ in that they involve local departures from the overall speaking rate. Note that the time distribution of non-intentional pauses (e.g., breathing pauses) can also be controlled intentionally, at least under some circumstances. For instance, in reading aloud, breathing pauses have been observed to occur at clause boundaries, whereas in spontaneous speech, only one third of them occurred at these locations (Goldman-Eisler, 1968, quoted by Butterworth, 1980). It may be that controlled departures from the basic speaking rate are programmed timing properties of speech like rubato in music. Or alternatively, they may result from the relaxation of the speech motor system, somehow linked with the non-continuous transmission of higher-order linguistic information packages. In this case one would be faced with emergent rather than truly programmed timing properties of speech.

Stress patterns and segment duration

The syllables of a word are usually not all pronounced with uniform temporal and dynamic characteristics. In many languages, one syllable, the *stressed* syllable, is more prominent than others. Stressed syllables are produced with greater intensity and/or longer durations than nonstressed syllables. Stressed and nonstressed syllables may also differ in the phonetic quality of vowel they contain. In some languages, the location of the stressed syllable within a word is fixed; in other languages, the location of the stressed syllable can vary and different stress locations can also be used to distinguish between semantically different words with otherwise identical forms. (e.g., in Russian, English, etc.). It has been argued that in these lexical-stress languages the word's stress pattern is an integral part of the word's phonological representation in the mental lexicon (Cutler, 1986). Whether or not a word's lexical stress pattern plays a role in its recognition has been a controversial issue (Cutler, 1984; 1986; Cutler & Clifton, 1984); mis-stressing, however, is known to inhibit spoken word recognition (see Cutler, 1986). Correct word stressing is thus an important constraint on speech production. One may ask the question as to whether the stress pattern of a word is coded in terms of *temporal features* (i.e., the distribution of relative durations), or whether the actual timing of the segments and syllables within the word may not result from specifying a quality of the articulative gesture (e.g., distribution of forces), as Shaffer (1984) has suggested. There is some consistency between the latter suggestion and the fact that the differences which exist between the so-called „inherent" durations of various speech segments are also closely correlated with the biomechanical properties of the articulators that generate them (see Klatt, 1976). There nevertheless exist reasons for believing that timing goals may also be specified in the speech motor program. For example, the speakers of a language must have some internal representation of the range of *permissible deviations* from the „inherent" segment duration when the segment occurs in a stressed or unstressed syllable. Furthermore, temporal cues are of overwhelming importance in speech perception, and therefore, the production systems are likely to be „attuned" to this fact. For instance, the perceived ratio between the durations of the stressed and unstressed syllables of two-syllable words is a more reliable cue to stress location than the ratio between their perceived intensities (Fry, 1955). In addition, there are languages in which the duration of vowels as well as that of consonants may serve to differentiate semantically between otherwise identical couples of words. This is the case in Hungarian, where differential timing patterns of this kind (generally having duration ratios of 1:2 between the short and long version of the same segment) exist quite independently of the dynamic stress (Molnar, 1988). Timing characteristics of this kind are likely to be preselected in a motor program.

Higher order rhythms

The pattern of accentuation of a *sentence* does not simply arise from the lexical stress patterns of the words which constitute the sentence. In any sentence, at least one word receives greater focus or emphasis than the others; consequently, the normal stress pattern of that word is enhanced, and the stress pattern of neighboring words may be remodeled. The remodeling of a word's stress pattern in the context of a sentence, or because of the number of syllables in the word, often fulfills the requirement to produce rhythmic alternation between strong and weak, longer and shorter syllables (see, e.g., Kelly & Bock, 1988). Higher order timing characteristics of this kind have been described in the case of various languages. In Hungarian, for instance, the emphasized word in a sentence receives a strong accent on its first syllable; in longer words, secondary accents are assigned to the odd-numbered syllables. The second and last syllables are always unstressed (Molnar, 1988). These timing characteristics are probably specified in the speech motor program.

Speech perception is efficiently attuned to periodicities, even in neonates (as mentioned by Cutler & Mehler, 1993); the non-occurrence of an expected rhythm may cause recognition to be momentarily disrupted (Cutler, 1991). There has been some debate as to whether in „stress-timed" languages such as English, strong stress recurs in normal speech with a strict periodicity (isochrony). Although the available evidence does not support this hypothesis at the most general level (Cutler, 1991), some data have shown that isochrony does exist in limited subunits of normal speech, called „narrow rhythm unit." The duration of these units tends to be independent of the number of phones it involves (Jassem et al., 1984), as if a time scale was allocated to the group of phones as a whole.

Keyboarding

Users of a keyboard (typewriting machine, computer, piano) often show astonishing skill in producing fast and fluent motor output. Here again, the temporal characteristics of the motor output can be examined at two levels: at the level of fairly continuous movements and at the level of the discrete events produced by movements. Since in keyboard tasks, the terminal motor output consists of easily identifiable discrete events, i.e., keypressing, studies on timing have tended to deal almost exclusively with the analysis of the statistical properties (mean, variance, sequential dependencies) of the inter-key intervals. Far less attention has been paid to the temporal organization of the movement trajectories themselves. In a study of this kind, recordings of the hand and finger movements occurring during fast typewriting showed, for instance, that there existed a high degree of temporal overlap (coarticulation) between the movements of the various effectors. They also showed that the temporal variability of the movement onsets was much greater than the temporal variability of the inter-key intervals (Gentner et al., 1980). As Shaffer (1984) has pointed out, the timing of movement onsets is subordinate to the timing of behaviorally meaningful events, that is, the key-pressing.

In many keyboard tasks, the timing of the events involves only moderately demanding timing requirements. In typewriting and in various laboratory tasks for instance, the output events have to be ordered correctly and produced fluently at a high speed; their exact moment of occurrence is not constrained however. In contrast, in piano playing, the musical events (and hence, the key pressing) must occur at fairly strictly determined points in time. In other word, the event timing must in this case fulfill real-time requirements.

Event timing in a non-musical context

Like speech events, keypressing performances show some *temporal variability*; some of the variations are purely random, while others are quite systematic even in cases where the task does not require any definite time shape to be performed. A very simple keyboard task consists of requiring a subject to tap repeatedly with the same finger on the same key, either at a predetermined rate or „as fast as possible" (Garcia-Colera & Semjen, 1987). In short strings of three to eight such taps, a characteristic lengthening of the initial and terminal intervals can be observed (Figure 2 A, p. 33), which is reminiscent of the pauses that occur at syntactic boundaries in speech. In very fast tapping sequences, the terminal lengthening is absent in three-tap sequences (i.e., in sequences with two inter-tap intervals); it develops over several intervals in longer sequences (Figure 2 B, p. 33). Although the lengthened intervals might function as boundary-markers, there seem to be few grounds justifying a programming account of such timing variations.

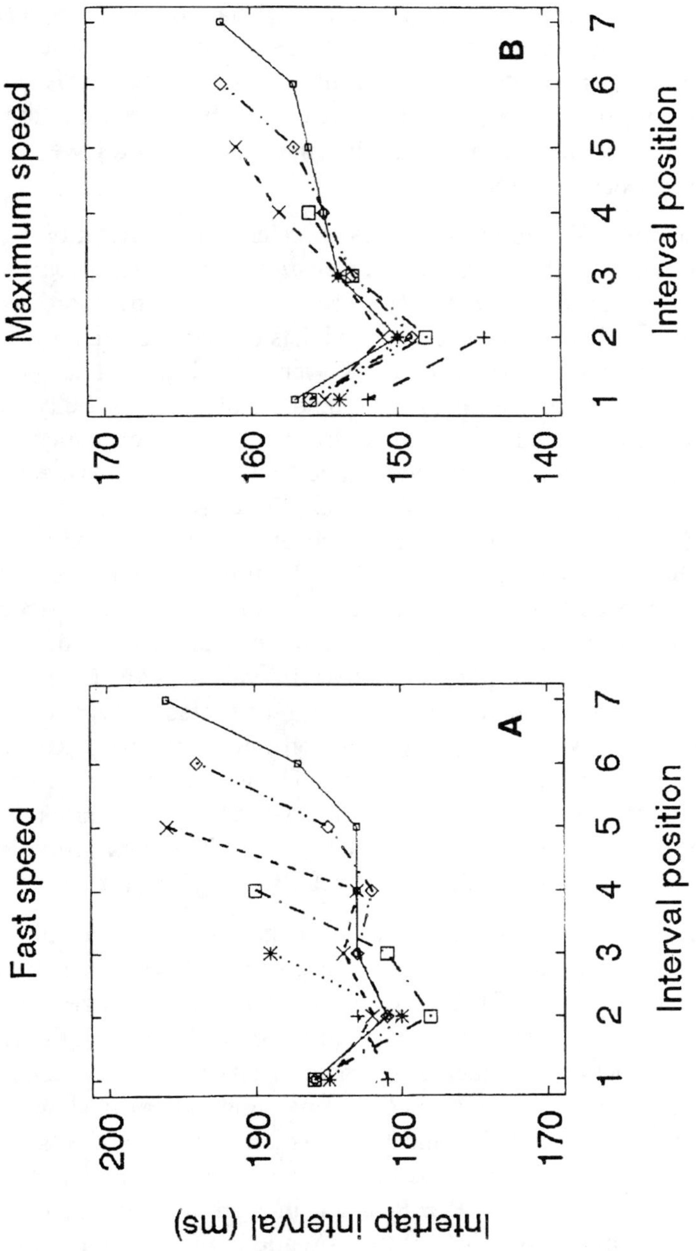

Figure 2: Mean intertap intervals (ms) at successive positions in tapping sequences of three to eight taps. (A): tapping at a predetermined rate. (B): tapping as fast as possible (after Garcia-Colera & Semjen, 1987).

It seems more likely that they result from the oscillatory dynamics of the effector system. Interestingly, a similar pattern has been observed in the timing of equal-note groups in piano playing (Shaffer et al., 1985). In these observations, the beat interval was to be subdivided into four to eight equal note durations playing all the notes with the right hand. In the actual performances, the first and last notes were played with a lengthened duration.

Contrary to the above mentioned acceleration-deceleration pattern, some inter-key patterns show *deceleration first* (i.e., progressively lengthening intervals) *followed by acceleration*. A pattern of this kind was observed when skilled typists typed lists of words containing 5 to 11 characters (Ostry, 1983). The words were typed as fast as possible. Care was taken to equate, over the whole set of words to be typed, the number of within-hand transitions (successive characters typed by fingers of the same hand) and between-hand transitions (successive characters typed with alternating hands) because these factors are known to influence the typing rate. The deceleration of execution during the initial part of the sequence was attributed to the time-consuming processing operations needed to organize the later part of the sequence. According to this hypothesis, the slowing down of the execution was due to processing operations taking place concurrently with the execution. That is, the timing was taken to reflect information processing. An alternative explication however is that the production rate may have been directly controlled, i.e., the subjects may have modulated their typing rate so as to allow for the completion of the necessary processing operations. From this point of view, the timing was directly controlled; appropriate timing created the conditions for appropriate sequence organization.

Support for the latter kind of interpretation was obtained in a study in which subjects produced finger-tapping sequences with their right and left hands (Semjen, 1992). The sequences consisted of finger taps executed either with the same finger (repetition) or with different fingers (alternation). A definite number of repeated or alternating taps constituted distinct sub-units of the sequence. In one situation, each subunit contained either two repeated elements or two alternating elements and the sequence consisted of eight taps in all. (A sequence of this sort can be written RR/LR/LL/RL/, where R and L indicate a right and left index finger-tap, respectively.) In another situation, the subunits contained four elements and the sequence consisted of 16 taps in all (e.g., RRRR/LRLR/LLLL/RLRL). The subjects were requested to perform the sequences fluently, as fast as possible, 6 times without any interruption. The results showed that the sequences having larger subunits were performed at a significantly faster rate that the sequences made up of smaller subunits. This was taken to indicate that the subjects adapted their rate of execution to the conditions of on-line sequence organization. When the subunits were larger, the organization of the forthcoming subunit could be completed in time even if the execution rate was fast. Conversely, when the subunits were smaller, the processing of the forthcoming subsequence could be completed only if the execution rate was somewhat slower. A further finding was that the pattern of inter-tap intervals showed within-subject consistency over successive renderings of the sequence. There were, however, considerable between-subject variations in the patterning of the inter-tap intervals. Some subjects introduced a longer interval („pausing“) between the subunits, and other subjects lengthened the first interval within the subunits. One subject even completely changed the timing profile of the same sequence from one session to the next. All in all, these findings show that different timing profiles can be superimposed on the same sequence, depending on individual movement organization strategies and on the context.

Unlike the above laboratory task, typewriting constitutes a highly complex cognitive and motor skill. After months and years of practice, typists acquire the ability to produce very fast and fluent motor output which nevertheless shows non-random fluctuations between the inter-key times. A large proportion of these timing variations reflect bio-mechanical constraints such as typing with the same finger, with different fingers of the same hand, or with alternating hands, or making movements between variably distant rows of letters on the keyboard. Typists develop „motor solutions“ to cope, in an anticipatory manner, with at least some of these contingencies by creating overlapping movement trajectories, which in turn impose a definite shape on the inter-key time patterns (Shaffer, 1984).

It has been suggested that in typewriting, words are motor programming units because they are often typed with consistent inter-key time patterns. These timing patterns were thought to reflect centrally stored *timing engrams*. At the time of actual production of a word, a so-called multiplicative rate parameter might scale the *relative timing pattern* stored in the engram (Terzuolo & Viviani, 1980). However attractive these suggestions may be, they should be considered with extreme caution. Word timing has been found to depend both on left context and right context (Shaffer, 1978); these contextual effects often cross word boundaries. Extensive statistical analysis did not support the notion that the relative timing of keystrokes in many words may be invariant (Gentner, 1982). Finally, the word timing patterns are subject-specific, just like the walking and running rhythms which often help us to recognize a person moving in the distance. The degree of stability of the timing „signature“ of a typed word may simply be an emergent characteristic which depends on the dynamics of the motor gestures involved in the word's actual production (Shaffer, 1984).

Internal timekeeper

Although in skilled typing, only near periodicity and fast and fluent execution are required, more stringent criteria have to be applied, for instance, in timing musical events. This requirement is even more obvious when several people are playing together: they need some shared internal representation of time, and this representation can be provided either by an internal clock or timekeeper, or by hierarchically arranged timekeepers (Vorberg & Wing, in press). Wing & Kristofferson (1973) have modeled some basic properties of simple motor sequences timed by a regular, self-sustaining internal timekeeper (Figure 3 A, p. 38). These authors assumed that the timekeeper can be set to generate time intervals having the desired mean duration (C); the mean interval is generated with some random variability. At the end of each timekeeper period, the timekeeper emits a trigger pulse which initiates a series of covert motor processes leading to an overt motor action. The motor processes introduce a delay between the occurrence of the trigger pulse and the occurrence of the observable response; this delay has a mean D and random variability. The variance of the timekeeper and the variance of the motor delay are taken to be uncorrelated. Given these assumptions, the total variance of the intervals between observable motor events can be decomposed into timekeeper variance and motor delay variance. The model predicts that (a) the timekeeper variance will depend on the length of the timekeeper period, whereas the motor delay variance will not; (b) adjacent inter-response intervals will be negatively correlated; and (c) this correlation will be bounded ($-.5 < r < 0$). Although these predictions have been confirmed experimentally (Wing, 1980), the strong emphasis placed in the model on the timing of movement *onsets* makes it questionable. We have noted that the timing of movement onsets is generally more variable than that of the events generated by the movements; and we have mentioned what could be called a general principle of movement organization in time, namely, that the timing of movement onsets and movement trajectories is subordinated to the timing of the target events (Shaffer, 1984). The problem of „endpoint programming“ has been investigated recently in a series of experiments on the interaction between timing control and force control in periodic tapping (Billon et al., in press; Billon & Semjen, 1995). In one experiment (Billon & Semjen, 1995), two groups of subjects, consisting respectively of musicians and non-musicians, were requested to produce repetitive finger taps on a response key. Sequences of 55 taps were performed in each trial either synchronously with a metronome, or simply continuing the rhythm of a metronome which was turned off when the subject's first tap occurred. The subjects were given an additional task consisting of accentuating every fifth tap. Accentuation is known to alter the regular timing of taps; the accentuated tap is generally followed by a lengthened interval (Semjen & Garcia-Colera, 1986). The trajectory of the tapping movements was recorded and inter-response intervals were determined between the onsets of the successive tapping downstrokes, and between the offsets of the successive downstrokes (i.e., between the successive finger taps on the response key). The aim was to attempt to answer the question as to whether (a) the presence of a metronome and/or the subjects' musical expertise might reduce or eliminate the lengthening of the interval after the accentuated tap, and (b) the timing of the taps (movement offsets) might be more regular than the timing of the downstroke initiations (movement onsets), as suggested by the „endpoint programming" hypothesis. The results showed that musicians and non-musicians both produced keytap intervals with virtually identical timing profiles. The interval after the accentuated tap was lengthened in both groups to the same extent, despite the presence of the metronome. Although the musicians' and non-musicians' groups did not differ in the timing of the goal events (taps), they differed dramatically in the way they organized and timed their tapping movements. In the non-musicians' group, the tapping downstrokes showed large, systematic time variations: the accentuated downstroke had a much shorter duration, and the subsequent downstroke, a much longer duration, than the other downstrokes. In contrast, musicians tried to equalize the time it took for them to complete the tapping downstrokes. How was it, then, that the two groups showed almost identical keytap timing? This was possible because the subjects in the non-musicians' group compensated in advance for the quick execution of the accentuated movement by systematically delaying its onset. Although the two groups produced different and quite variable patterns of movement onset timing, their keytap timing was therefore almost identical and less variable. Moreover, the same contrast appeared between the groups when the tapping was entirely self-paced, i.e., when no external metronome was present. These findings can be taken to indicate that the internal clock provides temporal goal points for meaningful events (Figure 3 B, p. 38) rather than a time-table for movement onsets.

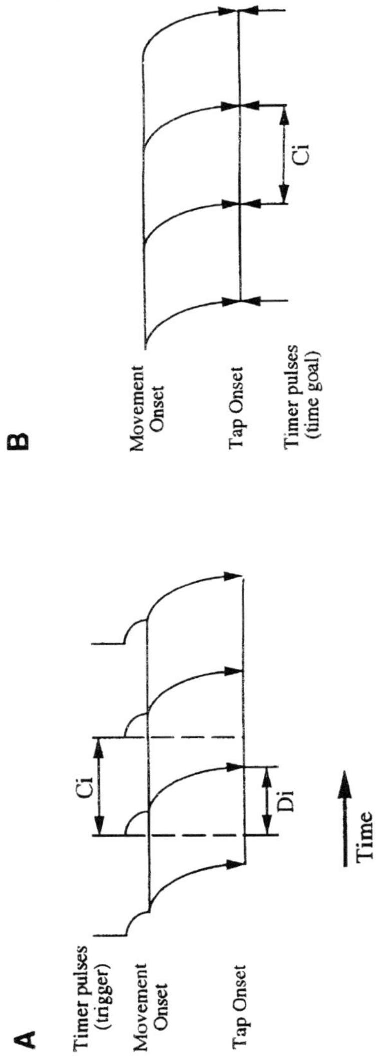

Figure 3: Models of clock-timing of periodic tapping.
(A): Trigger model of timing.[2]
(B): Endpoint model of timing.[3]

---

2 The output pulses of the timekeeper trigger motor processes (partly internal to the nervous system) which result in observable movements. The observed response events (e.g., taps) are recorded with a delay (D_i) relative to the occurrence of the trigger pulse. C_i refers to the preset period of the timekeeper (after Wing & Kristofferson, 1973).

3 The internal timekeeper provides temporal goal points at which target vents should occur.

The idea of endpoint timing contains the implicit notion that the time frame provided by the internal clock may make it possible to coordinate streams of independently unfolding events. For instance, a pianist playing a 3:4 polyrhythm, that is, subdividing the beat into three equal notes with one hand, and into four equal notes with the other hand, may take the internal beat as a moment at which the notes played by each hand should coincide. By modulating the rate of the internal timekeeper, it may be possible to combine rubato with polyrhythm, as Shaffer has shown (Shaffer 1981). It should be noted, however, that these kinds of temporal coordination require motor independence between the limbs. Between-hand temporal coordination is known to be easy, for average subjects, if the hands tap harmonically related periods, but it becomes very hard when producing non-harmonically related periods or polyrhythms (see Summers, 1990). This indicates that lower-level interactions between coupled oscillatory systems may override the cognitively defined timing goals.

Conclusion

In this chapter we have reviewed some empirical findings and some theoretical analyses which seemed to be relevant to the distinction between two levels of temporal organization in motor timing. These levels involve the organization and coordination of movements on the one hand, and the organization of the movement-produced events, on the other hand. One aspect of the coordination between concurrently unfolding movements is their relative timing which might emerge, we have argued, as the result of self-organizing processes in the motor systems. At this level, time might not be specified as a separate dimension; rather, it appears to be intrinsically bound up with the functioning of the systems. The temporal coordination of movement-produced events must often satisfy, on the other hand, sets of pre-specified instructions or temporal goals which are to some extent extrinsic to the executive systems (for an analysis of „extrinsic“ and „intrinsic“ timing, see Fowler, 1980). How the motor systems may learn to accept these timing representations (programs) as inputs, and to select and implement the corresponding motor procedures, is a matter which requires further research in the field of motor programming.

References

Abbs, J.H. & Connor, N.P. (1989). Motor coordination for functional human behavior: perspectives from a speech motor data base. In: S.A. Wallace (Ed.), *Perspectives on the coordination of movement* (pp.157-182). Amsterdam: North Holland.

Ballmer, T.T. (1978). The role of pauses and suprasegmentals in a grammar. Paper for *Pausological implications of speech production*. An interdisciplinary workshop, Kassel, Germany.

Billon, M., Semjen, A., & Stelmach, G.E. (in press). The timing effects of accent production in periodic finger-tapping sequences. *Journal of Motor Behavior*.

Billon, M., & Semjen, A. (1995). The timing effects of accent production in synchronization and continuation tasks performed by musicians and nonmusicians. *Psychological Research*, *58*, 206-217.

Butterworth, B. (1980). Evidence from pauses in speech. In: B. Butterworth (Ed.), *Language production 1* (pp. 155-176). New York: Academic Press.

Cooper, W.E. (1980). Syntactic-to-phonetic coding. In: B. Butterworth (Ed.), *Language production 1* (pp. 297-333). New York: Academic Press.

Cooper, W.E., Paccia, J.M., & Lapointe, S.G. (1978). Hierarchical coding in speech timing. *Cognitive Psychology*, *10*, 154-177.

Cutler, A. (1984). Stress and accent in language production and understanding. In: D. Gibbon & H. Richter (Eds.), Intonation, accent and rhythm (pp. 77-91). New York: Walter de Gruyter.

Cutler, A. (1986). Forbear is a homophone: lexical prosody does not constrain lexical access. *Language and Speech*, *29*, 201-220.

Cutler, A. (1991). Linguistic rhythm and speech segmentation. In: J. Sundberg, L. Nord, & Carlson, R. (Eds.), *Music, language, speech and brain. Proceedings of an international symposium* (pp. 157-166). Houndmills, Basingstoke: MacMillan Press.

Cutler, A. & Clifton, C. (1984). The use of prosodic information in word recognition. In: H. Bouma & D.G. Bouwhuis (Eds.), *Attention and Performance X. Control of language processes* (pp. 183-196). Hillsdale N.J.: Erlbaum.

Cutler, A. & Mehler, J. (1993). The periodicity bias. *Journal of Phonetics*, *21*, 103-108.

Dogil, G. (1984). Grammatical prerequisites to the analysis of speech style: fast/casual speech. In: D. Gibbon & H. Richter (Eds.), *Intonation, accent and rhythm* (pp. 91-119). New York: Walter de Gruyter.

Fowler, C.A. (1980). Coarticulation and theories of extrinsic timing. *Journal of Phonetics*, *8*, 113-133.

Fowler, C.A., Rubin, P., Remez, R.E., & Turvey, M.T. (1980). Implication for speech production of a general theory of action. In: B. Butterworth (Ed.), *Language production 1* (pp. 373-420). New York: Academic Press.

Fry, D.B. (1955). Duration and intensity as physical correlates of linguistic stress. *The Journal of the Acoustical Society of America*, *27*, 765-768.

Garcia-Colera, A. & Semjen, A. (1987). The organization of rapid movement sequences as a function of sequence length. *Acta Psychologica*, 66, 237-250

Gentner, D.R. (1982). Evidence against a central control model of timing in typing. *Journal of Experimental Psychology: Human Perception and Performance*, 6, 793-810.

Gentner, D.R., Grudin J., & Conway, E. (1980). *Finger movements in transcription typing*. Technical Report 8001. La Jolla, Calif., University of California at San Diego, Center for Human Information Processing.

Gracco, V.L. & Abbs, J.H. (1986). Variant and invariant characteristics of speech movement. *Experimental Brain Research*, *65*, 156-166.

Gracco, V.L. & Abbs, J.H. (1988). Central patterning of speech movements. *Experimental Brain Research*, *71*, 515-526.

Jassem, W., Hill, D.R., and Witten, I.H. (1984). Isochrony in English speech: its statistical validity and linguistic relevance. In: D. Gibbon & H. Richter (Eds.): *Intonation, accent and rhythm* (pp. 203-225). New York: Walter de Gruyter.

Kelly, M.H. & Bock, J.K. (1988). Stress in time. *Journal of Experimental Psychology: Human Perception and Performance*, *14*, 389-403.

Kelso, J.A.S. (1981). Contrasting perspectives on order and regulation in movement. In: J. Long & A. Baddeley (Eds.), *Attention and performance IX* (pp. 437-457). Hillsdale NJ: Lawrence Erlbaum.

Klatt,D.H. (1976). Linguistic use of segmental duration in English: acoustic and perceptual evidence. *Journal of the Acoustic Society of America*, *59*, 1208-1221.

Molnar, E. (1988). *Leiro magyar hangtan* (Descriptive Hungarian phonetics). Tankönyvkiado, Budapest.

Ostry, D.J.(1983) Determinants of interkey times in typing. In: W.E. Cooper (Ed.), *Cognitive aspects of skilled typewriting* (pp. 228-246). New York: Springer.

Perkell, J.S. (1980). Phonetic features and the physiology of speech production. In: B. Butterworth (Ed.), *Language production 1* (pp. 337-372). New York: Academic Press.

Semjen, A. (1992). Plan decoding and response timing during execution of movement sequences. *Acta Psychologica*, *79*, 255-273.

Semjen, A. & Garcia-Colera, A. (1986). Planning and timing of finger-tapping sequences with a stressed element. *Journal of Motor Behavior*, *18*, 287-322.

Shaffer, L.H. (1978). Timing in the motor programming of typing. *Quarterly Journal of Experimental Psychology*, *30*, 333-345.

Shaffer, L.H. (1981). Performances of Chopin, Bach and Bartok: Studies in motor programming. *Cognitive Psychology*, *13*, 327-376.

Shaffer, L.H. (1984). Motor programming in language production. In: H. Bouma and Don G. Bouwhuis (Eds.), *Attention and performance X. Control of language processes* (pp. 17-41). Hillsdale N.J.: Erlbaum.

Shaffer, L.H., Clarke E.F., & Todd, N.P. (1985). Meter and rhythm in piano playing. *Cognition*, *20*, 61-77.

Summers, J.J. (1990). Temporal constraints on concurrent task performance. In: G.R. Hammond (Ed.), *Cerebral control of speech and limb movements* (pp. 661-680). Amsterdam: North Holland.

Terzuolo, C.A. & Viviani, P. (1980). Determinants and characteristics of motor patterns used for typing. *Neuroscience*, *5*, 1085-1103.

Vorberg D. & Wing, A.M. (in press). Modeling variability and dependence in timing. In: H. Heuer & S.W. Keele (Eds.), *Handbook of perception and action. Vol 3. Motor Skills*. London: Academic Press.

Wing, A.M. (1980). The long and short of timing in response sequences. In G.E. Stelmach & J. Requin (Eds.), *Tutorials in motor behavior* (pp. 469-486). Amsterdam: North Holland.

Wing, A.M. & Kristofferson, A.B. (1973). Response delays and the timing of discrete motor responses. *Perception and Psychophysics*, *14*, 3-12.

Chapter 3: How prospective time shapes behavior

ALF C. ZIMMER

The famous Kantian a-priories of space and time (1781; 1975) have in common that they constrain and sometimes even determine human behavior but are rarely experienced consciously. The geometry of the perceived world not only allows us to identify objects and their spatial relations to each other but it also determines the position of the perceiver towards the perceived objects. Apparently, for the perceiver it does not constitute a problem that the geometry of the „objects as they are" differs from the geometry of the „objects as seen from a specific perspective" (Zimmer, 1995). Actually the complementarity of these geometries determines the properties of the perceived world. Only under rare circumstances the disparity of these geometries is realized, for instance in the Ames-room or when perceiving the world in a camera obscura (Pirenne, 1970). In the case of time the situation is similar: The most direct perception of time happens in the experience of „time gaps" (Reed, 1972), where suddenly an ongoing action becomes conscious and one realizes that for a period of time this action has been performed without the experience of elapsing time, an everyday experience. For instance, one is driving a car through a well known environment when suddenly an unexpected change occurs and one wonders how one has come to this location. Furthermore, similarly as in space perception the experience of time consists of two distinct metrics: the metric of continuous time which is the same for all possible events, and the metric of episodes which is defined only for the episodes in question and therefore is discontinuous.

The first attempt to unravel the complexities arising from the fact that perceived time differs from the conceptual time of physics, was Kant's distinction of the „concept of time" and the direct experience of time („the image of time") which he assumed as coupled by the cognitive process of the „schema". Postulating this schema, however, does not resolve the problems of different metrics in time. A further problem which is also not resolved is the problem of defining time because time is measured via the amount of uniform actions (e.g. the movements of the escapement and the pendulum in mechanical clocks or the number of drops in a clepsydra). On the other hand, these actions and their uniformity are characterized by the fact that they need equal time. This together with the dual metric of perceived time motivates the researcher to approach the problem of perceived times directly by analyzing actions in time. Actions happen to be early, to be late or exactly at the right time, they may be hasty or to slow, and they may be accelerating or slowing. In order to experience these properties of actions it is necessary that these actions are perceived „as they are", that is, with their own characteristic metric, and simultaneously as mapped upon a continuous stream of time, that is, comparable to the metric of the uniform time.

The relations between the experience of time, how subjective duration is determined by the on-going activity, has been analyzed by Henri Bergson in his Oxford lecture on change in 1911. He explicitly assumes that change is constituted of elementary units below which no change happens; that is in line with the assumption of elementary actions constituting what is experienced as time: „*We have to assume that any change and movement is absolutely indivisible.* Let us start with a movement. My hand is at position A and I move it to position B running through distance AB. I postulate that this movement from A to B is something essentially simple. We can undoubtedly say that it is possible to stop our hand during the movement from A to B, but then it would be no longer the same movement. Neither internally with our muscular sensations nor externally with our vision would we have the same perception" (p. 158 italics in the original, translation by the author) and „*There is change but below change there are no changing objects*" (p. 163, italics in the original, translation by the author).

The indivisibility of change and the fact that actions taking different amounts of time are equally 'essentially simple' imply that subjective time is not a function of 'objective time' but of actions. If the units of action were identical for all humans, subjective time would no longer be relative as postulated by Kant in his „Critique of Pure Reason" (1781; 1975). Bergson states: „...it would not imply that it is necessary to transcend time in order to avoid Zenon's paradoxa and to free our perception from the relativity as postulated by Kant ... to distance us from change (...), on the contrary it would be necessary again to perceive change and duration in their original changeability" (1960, p. 157, translation by the author).

Starting with Donders (1869) in the last century, the approach of chronometric analysis of cognitive processes has attempted to resolve exactly these problems by mapping actions directly upon the physically measured time. The underlying assumption of this approach is that mental processes consist of elementary processes which are uniform concerning their temporal characteristic, an often used analogon for this kind of model are neuronal sequences of different length. The more complex actions necessitate the activities of more levels which are linked by longer sequences of nerves. If this analogue holds, the differences in reaction times mirror the differences in the complexity of the underlying organization for the action. Behind this „strong" model of chronometric analysis lie assumptions about the structure of the brain as carrier or locus of cognitive processes, namely, the assumption of a strict and decomposable hierarchy (see Sternberg, 1969).

However, starting with the analyses by McCulloch & Pitts (1943) or Hebb (1949) it has become evident that at least locally there are no hierarchical and serial, but netlike parallel and distributed organizations (McClelland et al., 1986). The „strong" model can only be saved if these parallel processes are confined to modules and if these modules are hierarchically ordered (Fodor & Pylyshyn, 1988). The investigations by Mishkin & Ungerleider (1982) about the separability of space and object perception seemed to support this assumption. However, more recent investigations by Goodale & Milner (1992) indicate that these modules are working in parallel and usually form distributed networks but not hierarchies.

Instead of the untenable „strong" model of chronometric analysis a more „weak" alternative seems possible according to which the temporal characteristics of planning reveal if the intended actions are represented as units (modules) or as heterogeneous complexes. Such an approach offers a tentative solution for the problem of the relation between actions and subjective time: the subjective time for an action is equivalent to the set of constituting units. In this model the complexity of an action is not defined via physical time because measuring the reaction times simply serves as an indicator for the unity or complexity of the representations of the intended actions.

It is not necessary to assume that the unitary quality of a representation is static because processes of learning reveal that very complex trains of actions might be represented as a single unit for the expert. If something is represented as such a unit, partial processes which might have been independent or unitary in the beginning of the learning can no longer be addressed directly (see e.g. Zimmer,1990; Zimmer & Körndle, 1988).

The question concerning the temporal structure of motor actions must be supplemented by the question about the natural unit of motor behavior. This question provides two opposite answers. On the one hand, the opinion of Bernstein and the Gestalt-psychologists, which states that 'natural motor units' exist, and are reflected in the principles of self-organization (for a historical and thematic connection between Bernstein and the Berliner School of Gestalt-Psychology, see Körndle & Zimmer, 1994). Contrary to this opinion are the association psychological or behaviorist concepts based on the idea proposed by Adams (1971) and also, even though not as obviously represented by Schmidt (1975) and Shapiro & Schmidt (1982). According to this view, the complexity of motor actions or programs depends on the amount of simple actions, voluntarily controlled.

Following this comparison, it appears paradoxical at first glance that Adams (1971) refers directly to Bernstein. The precise analysis of the concrete experiments of Adams (1971) and also of Schmidt (1975) show, however, that despite the systems-oriented view of the open loop-theory and the schema-theory of Schmidt (1975), the choice of pointing or position tasks restricts actions to those, which are mechanical and therefore do not show self-organization processes for non-linear systems.

The 'natural motor units' need not be considered as inborn, but rather as a result of the interaction between the organism and its environment. Therefore learning processes can be assumed, which are implicitly and adaptively controlled by the general setting of the organism and environment and not necessarily by an explicitly defined aim. One can assume that - and this has been shown in the studies by Kelso (1990) - that such processes are best characterized as synergetic self-organization processes (see Haken, 1991). The central question is not which theoretical description best fits the facts, but rather are the basic conditions for self-organization processes given. That is, after an adaptive development only one or a few parameters control the emerged motor unit, while considerably less complex constituents of these motor units depend on the same number or even more parameters. As the results on imagining motor reacts - in this case the production of a script (Zimmer, 1982) - indicate the individual motor performances are not sufficiently defined by pictorial - static or dynamic - representations but additionally the kinesthetic and dynamic representation works as an effective cue for the memory of motor acts.

In the dynamic organization of motor units, segments can be found, which are clearly separated by singularities, e.g. change of direction, reduction in speed to zero etc. If one would assume that motor units are represented as lists of those segments (e.g., it is implied in the program theory) then those segments would have to be faster and more precise than the whole movement. If however one would assume that in the internal representation the motor unit is primary and is defined by only one or few degrees of freedom, then the motor unit should be initialized as fast, if not even faster than the segments of it.

The process of the evolution of motor units has undoubtedly something in common with the concept of overlearning, where - at least according to the behaviorist tradition - the effects are assumed as purely additive and not self-organized. However, since the phenomena of overlearning refer to the whole motor units as well as to the separate segments of these motor units, it would only be relevant for the problem of the characteristics of motor units if there is a dissociation in overlearning between the explicit acquisition of the whole movement and the implicit acquisition of the partial movements.

Seemingly, a connection can be constructed between the motor theory of Adams (1973) with its central processes of 'timing' and 'sequencing' and Kien's notion of characteristic time structures (see e.g. Schleidt & Kien, 1994). Schleidt & Kien and partially Pöppel (1990) assume that the necessary time for tasks of a specific complexity is independent from the number of activated motor constituents as defined physiologically, but depends on the number of action units. In contrast, the concepts of Adams (1971) or Schmidt (1975) imply that this time correlates with the number of the separate components (i.e., parameters to be set) or in the case of established schemata with the time between the start signal and the beginning of the motor unit. This time is described in the literature as preparation time (for the bio-psychological fundamentals see Requin, 1980; Brunia et al., 1985). The study of Sternberg et al. (1978) and the new experiments of Kornbrot (1989; 1991) apparently confirm the position according to which complex actions require longer preparation times.

A traditional approach to the identification of such motor units is the chronometric analysis (Sternberg, 1969). The preparation times for sequences of motor actions should become longer the more complex the sequence of motor actions is. For instance, it should take longer to react after the „start"-signal if one has to type a word consisting of five letters than if a word contains two letters. This effect would be independent from perceptual processes because the subjects know in advance which word is coupled with which signal. It becomes apparent that independent from the number of actions to be executed, the level of cognitive processing remains the same (see Figure 1, p. 51), namely, the skill level

(Rasmussen, 1986). Operations on the rule or knowledge level are not necessary to perform these tasks.

The increase in reaction times is explained in the framework of information processing. Namely, the preparation of any motor action takes a constant time and these preparation times combine; in this case the additive factor method (Sanders, 1980) can be applied to the analysis of motor behavior. The identification of motor units is possible if tasks are found in which there is a linear increase of reaction times depending on an increase in complexity. Sternberg et al. (1978) have been the first to apply this theoretical principle to motor actions in speaking (see also Gracco, 1990). Further investigations, for instance, of tapping (Canic & Franks, 1989), typing (Gentner, 1987; Kornbrot, 1989), or grasping (Green, 1982), have lead to interpretable results. However, the attempt to identify the motor units in writing has given rise to contradictory results. The data of some experiments support the information processing model by showing a linear, albeit slight, increase of reaction times with an increase in the number of letters, in most experiments. However, no increase whatsoever has been found (see Hulstijn et al., 1983). The analysis of the different complexity of letters equally fails to produce consistent results in line with the assumption of information processing.

If letters or numerals were unitary actions, it should take longer to initialize writing the longer the sequence of symbols in the task is. If, however, the units of actions are lines, circles or segments of circles as Gibson & Levin (1975) suggest, the number of these constituents of letters should determine the preparation time. Neither assumption about the representation of handwritten symbols seems to hold because, as already said, in most experiments on handwriting one finds no differences in preparation times depending on the number or complexity of symbols in handwriting. The question arises why is handwriting so as compared to typing, pushing buttons, or speaking.

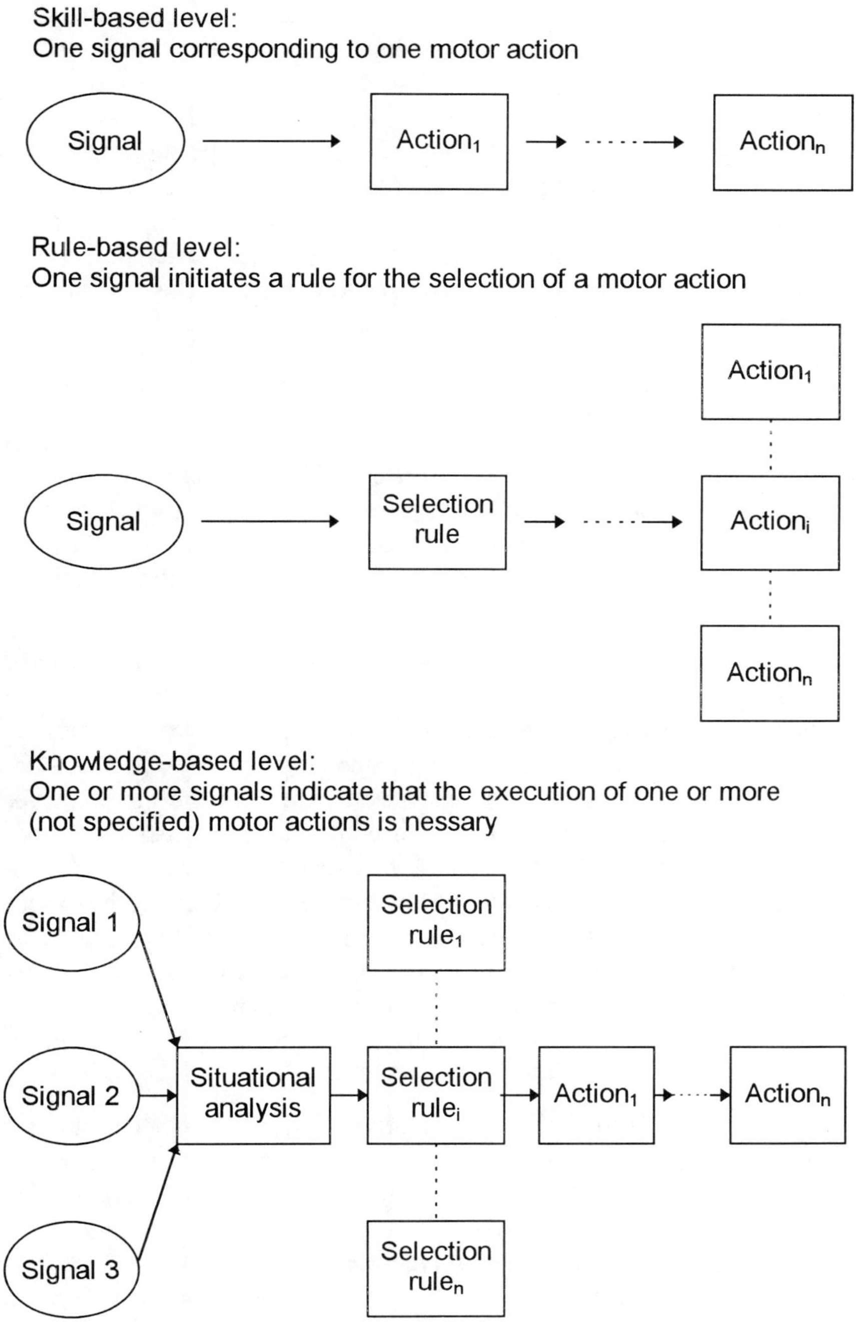

Figure 1: Levels of processing in the selection of actions (after Rasmussen, 1986)

One possible cause for these differences might be that subjects allocate their time in writing differently: Some, „the information processing“ subjects, take their time in planning, that is, more time is expended as we increase the complexity or number of symbols. Also, they execute the task ballistically, that is, when having started to write the subjects do not need to control or plan ahead anymore. Other subjects, the „opportunistic planners“, start the task immediately, that is, they do not show a latency which lengthens the more actions in the motor buffer have to be brought in sequence. However, when executing the task the time per letter increases proportionally with the number of letters to be executed, that is, the rate effect mirrors the amount of advanced planning in parallel to the execution of writing (Carcia-Colera & Semjen, 1988).

This has led to the question, what distinguishes motor behavior in writing from other repetitive motor actions. One possibility would be a different structural organization (as compared to a motor behavior in speaking). Another reason might be that different methods of writing exist and therefore averaging data over subjects does not result in a simple linear model.

Experiments about the time course in planning and executing motor actions point to two different directions: (i) There are experiments which - albeit under very restrictive conditions - support the notion of preparation times relative to the number and/or complexity of consecutive actions. The latency effect is explained by a search in the motor buffer which takes longer the more elements are in it and a „ballistical“ execution of motor actions after terminating this search. (ii) Other experimental data do not show this kind of latency but a „rate effect“ in the execution of the motor actions, that is, the more motor actions that have to be executed, the longer it takes to execute every single motor action. This can be explained by a process of opportunistic planning in which consecutive motor actions and monitoring the sequence while executing a motor action.

In order to illustrate how qualitatively different subjects allocate their time in preparing to act, data for five subjects are shown from an experiment where subjects had to write up to 5 numerals under time pressure (Figure 2, p. 53). Each data point represents the average of 20 reaction times. Not only that in an analysis of variance the subjects differ significantly, but there is also a significant interaction between the subjects and the number of numerals to be written; that is, subjects follow qualitatively different allocation strategies depending on the task at hand. Subject „BS“, for instance, becomes faster the more ciphers have to be written down, subject „PM“ shows a flat curve, and subject „AS“

shows the expected increase for the first three letters but then the curve declines again. Different allocation times in writing do not have an influence on the execution times as shown in Figure 3 (p. 54) where the quadratic form of the curve shows that the rate effect and the data for all subjects are practically identical.

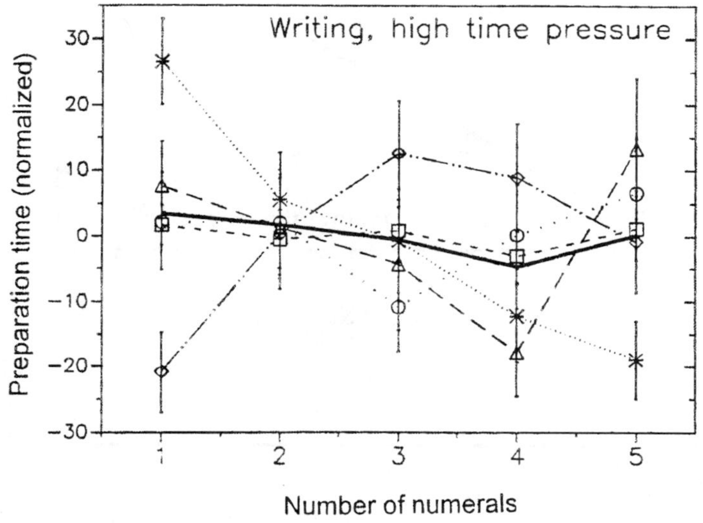

Figure 2: Normalized preparation times for subjects SM (O), AK (□), PS (Δ), RF (◊), RW (∗). The bold line indicates the averages and the bars the standard deviations.

A repetition of these experiments where subjects did not have to react as fast as possible, reveals that the quadratic form in Figure 3 (p. 54) is not coincidental but that subjects in writing really allocate their time opportunistically. In parallel to executing one letter they plan already the other letters: the rate effect.

In conclusion, the strong differences in the chronometric analysis between tasks which all increase with the number of required actions and instructions (*with* versus *without* time pressure) indicate that in regard to motor actions humans do not have an internal clock, that is, for them time is not defined independently from the task and the situation.

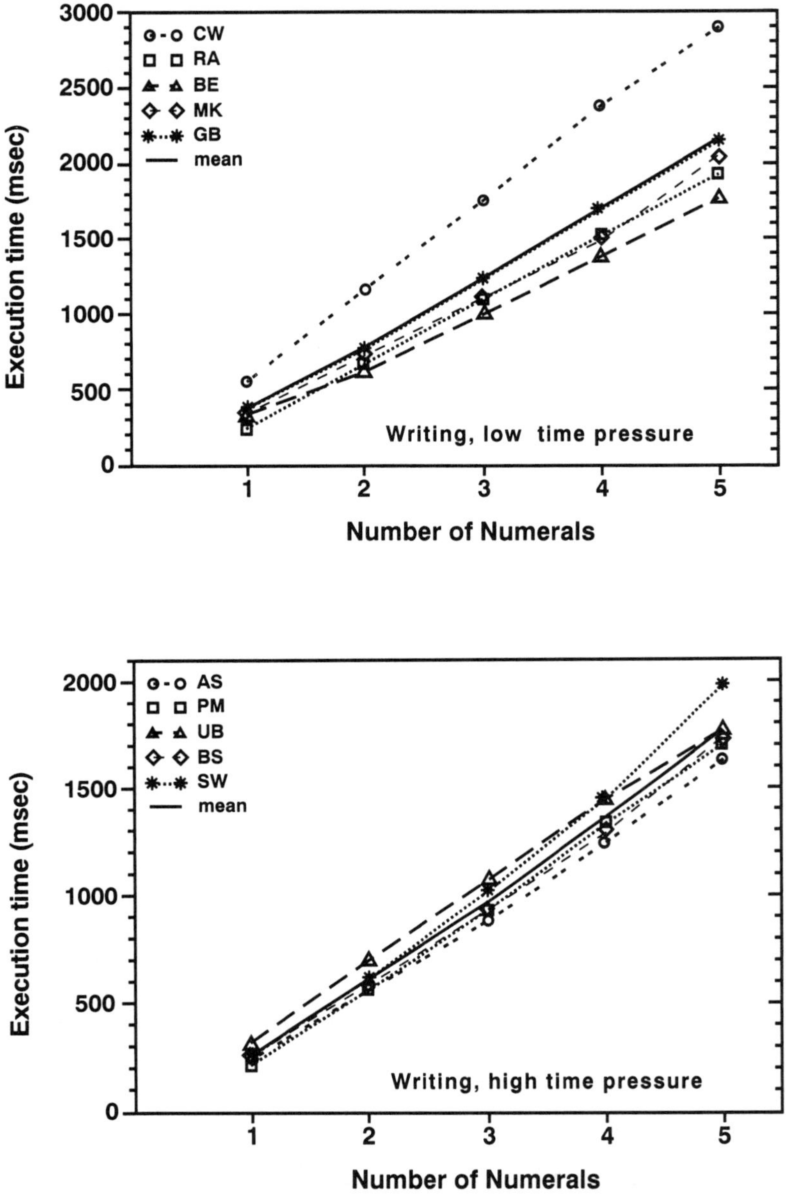

Figure 3: The execution times for writing 1 - 5 numerals. The trend is significantly quadratic for the average time as well as for 3 out of 5 subjects

A special case for this differential allocation of time is music where Grepmair & Zimmer (in preparation) have investigated if experts and novices in music (expert versus lay musicians) have different frames of reference for musical tempi. The surprising result is that amateur musicians are much more consistent in calibrating the musical tempi to beats on the metronome than experts. Professional pianists only when playing simple parts of a sonatina by Clementi (op. 36 Nr. 1 „allegro“, see Figure 4, p. 54) used a tempo as defined by metronome beats. However, a closer analysis of more complex parts reveals that some musicians slow down immediately before the complex parts but execute these according to the „objective“ tempo, while others slow down the execution of the complex part, that is, they use the technique of „rubato.“

Figure 4: Notation of the sonatine by M. Clementi, Op. 36 Nr. 1

As Crossley-Holland (1980, p. 745) remarks: „Such modifications of tempo, known as *tempo rubato - i.e.*, 'robbed time' - are part of the music's character. Rubato needs the framework of an inflexible beat from which it can depart and to which it must return."

For the listener, the two different techniques give rise to different impressions of tempo: in the first case a variation in tempo is experienced, namely, an increase of tempo in the complex parts. In the second case, the tempo is perceived as remaining equal for the entire piece. If one draws analogies to the perception of space, these experiences of time become very plausible, because in the perception of distances one observes the „clutter" effect, that is, the more discriminable objects lie between two distinct points the longer this distance is perceived, as apparent in the Oppel-Hering illusion. In time perception this implies that a constant ratio of frequency and or complexity of events to time constitutes the frame of reference, deviations of this ratio from a constant value give rise to the impression of variability in the flow of time.

From the theoretical considerations as well as from the experimental results the following tentative conclusion concerning the relation between action and perceived time seems plausible. Not only does the prospective time, that is, the time budget, shape future actions but the intrinsic structure of these actions also shapes the perceived time course. A unitary but complex action consisting of many constituents arranged in an intricate manner is experienced as shorter in time than a sequence of many simple actions, even if the overall number of constituents is the same.

Gerardo Dottori, a futuristic painter has depicted the complex texture of time and action in his painting 'speed' of 1925 (see Figure 5). Perhaps this picture captures best what is implied in the psychological aspect of time.

Summary

The perspective of research on the perception of time is usually retrospective,[1] that is, a sequence of actions is performed or a sequence of events is observed and afterwards the amount of elapsed time is estimated or the experience of elapsed time is recalled. These estimates and recollections usually correlate with the complexity or density of the sequence of events in time.

---

[1] However, see Block & Zakay, this volume, chapter 9.

Figure 5: Gerardo Dottori (1925) Velocitá, (tre tempi). Partenza

As a starting point of our investigations we have switched the perspective from retrospective to prospective, that is, how do the expectations about actions to be done influence our allocation of time. Time in this context is regarded as a scarce commodity which is spent cautiously and in relation to the complexity of the task at hand.

Different fields of action have been investigated: (i) writing signs, (ii) uttering words, and (iii) producing melodies, all on differing levels of complexity. In order to cope with this complexity two different strategies in allocating time are possible: (i) *the allocation of a complexity-dependent amount of time before starting* the entire sequence of actions. That is, reserving so much time after receiving the starting information that all relevant representations of motor actions can be activated and brought into the correct sequence with the subsequent motor activity in ballistic form, or (ii) *the quasi-opportunistic allocation of time* immediately before the single actions combined with a critical slowing down during parts of higher complexity.

The data of individual subjects reveal that they resolutely adhere to their - idiosyncratic - strategy of allocating time and only rarely do clear cut cases of strategy (i) occur. Mostly subjects exhibit variants of strategy (ii). Finally, experiments with more global motor behavior reveal that the actual timing of actions and the allocation of time aim at a smooth, „unjerky" interaction with the physical environment. „Jerky" motor behavior seems to occur only when there are abrupt changes in the environment or if the behavior is entrapped in a singularity.

References

Adams, J.A. (1971). A closed-loop theory of motor learning. *Journal of Motor Behavior, 3*, 111-149.

Bergson, H. (1911; 1960). La perception du changement. In H. Bergson (Ed.), *La pensée et le mouvant* (pp. 143-176). Paris: Presses Universitaires de France.

Brunia, C.H.M., Haagh, S.A. & Schirs, J.G. (1985). Waiting to respond: Electrophysiological measurements in man during preparation for a voluntary movement. In H. Heuer, U. Kleinbeck & K.-H. Schmidt (Eds.), *Motor behavior: programming, control, and acquisition* (pp. 35-78). Berlin: Springer.

Canic, M.J. & Franks, I.M. (1989). Response preparation and latency in patterns of tapping movements. *Human Movement Science, 8*, 123-139.

Crossley-Holland, P. (1980). Musical rhythm. *Enzyklopaedia Britannica. Macropaedia: Knowledge in Depth, 12*, 744-746.

Donders, F.C. (1869; edition 1969). On the speed of mental processes. *Acta Psychologia, 30*, 412-431.

Fodor, J.A. & Pylyshyn, Z.W. (1988). Connectionism and cognitive architecture: a critical analysis. *Cognition, 28*, 3-71.

Garcia-Colera, A. & Semjen, A. (1988). Distributed planning of movement sequences. *Journal of Motor Behavior, 3*, 341-367.

Gentner, D.R. (1982). Evidence against a central control model of timing in typing. *Journal of Experimental Psychology, 8*, 793-810.

Gentner, D.R. (1987). Timing of skilled motor performance: tests of the proportional duration model. *Psychological Review, 2*, 255-276.

Gibson, E.J. & Levin, H. (1989). *Die Psychologie des Lesens*. Frankfurt: Fischer.

Goodale, M.A. & Milner, A.D. (1992). Separate visual pathways for perception and action. *Trends in Neuroscience*, *15*, 20-25.

Gracco, V.L. (1990). Characteristics of speech as a motor control system. In G.E. Hammond (Ed.), *Cerebral control of speech and limb movements* (pp. 3-28). Amsterdam: Elsevier.

Green, P.H. (1982). Why is it easy to control your arms? *Journal of Motor Behavior*, *4*, 260-286.

Haken, H. (1991). *The synergetic computers and cognition*. Berlin: Springer.

Hebb, D.O. (1949). *The organization of behavior*. New York: Wiley.

Hulstijn, W. & van Galen, G.P. (1983). Programming in handwriting: reaction time and movement time as a function of sequence length. *Acta Psychologica*, *54*, 23-49.

Hulstijn, W. & van Galen, G.P. (1988). Levels of motor programming in writing familiar and unfamiliar symbols. In A.M. Colley & J.R. Beech (Eds.), *Cognition and action in skilled behaviour* (pp. 65-85). Amsterdam: Elsevier.

Kant, I. (1781; 1975). Kritik der reinen Vernunft. In W. Weischedel (Ed.), *Kant - Werke in 6 Bänden*. Darmstadt: Wissenschaftliche Buchgesellschaft.

Kelso, J.A.S. (1990). Phase transitions: foundations of behavior. In H. Haken & M. Stadler (Eds.), *Synergetics of cognition* (pp.249-268). Berlin: Springer.

Körndle, H. & Zimmer, A. (1994). A gestalt theoretic account for the coordination of perception and action. *Journal of Philosophical Psychology*, *2*, 249-265.

Kornbrot, D.E. (1989). Organisation of keying skills: the effect of motor complexity and number of units. *Acta Psychologica*, *70*, 19-41.

Kornbrot, D.E.. (1991). Levels of organisation and the planning of movement sequences: sequential evidence. In J. Requin, G.E. Stelmach (Eds.), *Tutorials in motor neuroscience. NATO ASI Series D: Behavioral and social sciences*, *62* (pp. 385-396). Dordrecht, Netherlands: Kluwer Academic.

McCollouch, W.S. & Pitts, W.H. (1943). A logical calculus of the ideas immanent in nervous activity. *Bulletin of Mathematical Biophysics*, *5*, 115-133.

McClelland, J.L., Rumelhart, D.E., & Hinton, G.E. (1986). The appeal of parallel distributed processing. In D.E. Rumelhart, J.L Mc Clelland & the PDP Research Group (Eds.), *Parallel distributed processing. Vol. 1* (pp. 3-44). Cambridge, MA: MIT Press.

Mishkin, M. & Ungerleider, L.G. (1982). Contribution of striate inputs to the visuospatial functions of parieto-preoccipital cortex in monkeys. *Journal of Brain and Behavioral Sciences, 6*, 57-77.

Pirenne, M.H. (1970). *Optics, painting and photography*. Cambridge: Cambridge University Press.

Pöppel, E., Ruhnau, E., Schill, K., & von Steinbuechel, N. (1990). A hypothesis concerning timing in the brain. In H. Haken, M. Stadler (Eds.), *Synergetics of cognition* (pp. 144-149). Berlin: Springer.

Rasmussen, J. (1986). *Information processing and human-machine interaction*. Amsterdam: North-Holland.

Reed, G. (1972). *The psychology of anomalous experience*. London: Hutchinson University Library.

Requin, J. (1980). Toward a psychobiology of preparation for action. In G.E. Stelmach & J. Requin (Eds.), *Tutorials in motor behavior* (pp. 373-398). Amsterdam: North-Holland.

Sanders, A.F. (1980). Stage analysis of reaction processes. In G.E. Stelmach & J. Requin (Eds.). *Tutorials in motor behavior* (pp. 331-354). Amsterdam: North-Holland.

Shapiro, D.C. & Schmidt, R.A. (1982). The schema theory: recent evidence and developmental implications. In: J.A.S. Kelso & J.E. Clark (Eds.), *The development of movement control and coordination* (pp. 113-150). New York: Wiley.

Schmidt, R.A. (1975). A schema theory of discrete motor skill learning. *Psychological Review, 82*, 225-260.

Schleidt, M. & Kien, J. (1994) Temporal organisation of action in baboons. *Brain Behavior and Evolution*, 44(2), 101-107.

Sternberg, S., Monsell, S., Knoll, R.L., & Wright, C.E. (1978). The latency and duration of rapid movement sequences: comparisons of speech and typewriting. In G.E. Stelmach (Ed.), *Information processing in motor control and learning* (pp. 117-152). New York: Academic Press.

Zimmer, A. (1982). Do we see what makes our script characteristic - or do we only feel it? Modes of sensory control in handwriting. *Psychological Research*, *44*, 165-174.

Zimmer, A. (1990). Der Erwerb komplexer motorischer Fertigkeiten. In C.Graf Hoyos & B. Zimolong (Eds.), *Ingenieurpsychologie* (pp. 148-177). Göttingen: Hogrefe.

Zimmer, A. (1995). Höhere Stufen der Wahrnehmung? In G. Kebeck (Ed..), *Gestalttheorie als Forschungsperspektive. Festschrift zur Emeritierung von Manfred Sader* (pp. 25-73). Münster: LIT-Verlag.

Zimmer, A. & Körndle, H. (1988). A model for hierarchically ordered schemata in the control of skilled motor action. *Gestalt Theory*, *10*, 85-102.

Part III:

Time perception and memory

Chapter 4: Time perception from a psychophysicist's perspective*

HANNES EISLER

It is commonplace to state that time perception is a broad subject and that a psychophysical approach is not always the most suitable way to tackle its problems. What I am interested in is the measurement of subjective (experienced) duration of short (less than a minute) time intervals. I shall try to be a narrow-minded psychophysicist, shying away from problems outside this limited frame of reference.

Brief summary of psychophysical experimentation

In psychophysical experiments usually some kind of standard is presented. If the standard consists of a duration, the subjects' task may be to identify it (for instance identify the duration as long or short), estimate it (for instance in seconds), reproduce it, or compare it to another duration („longer," „shorter"). Reproduction is a special case of ratio-setting. When the standard is given as a number (of seconds, for instance), the subject is to produce a duration corresponding to the number. The durations, whether standards or responses have, of course, to be indicated in some way, for example as sound, light, touch intervals, empty intervals limited by clicks, or as intervals filled with some kind of non-temporal task. (For a more thorough description, see Zakay, 1990).

Misunderstandings and bad habits

Both in literature studies of time perception and when discussing with fellow scientists I have occasionally noticed a certain naïveté regarding psychophysics. Here are some of the traps into which researchers sometimes fall.

---

* This investigation was supported by the Bank of Sweden Tercentenary Foundation, Project No. 95-0309:01.

Reporting of data

Deviation from standards without direction given

Response deviations from the standard are given without indicating the direction, for instance as a mere percentage „error“ (Pumpian-Mindlin, 1935) or just the statistical significance (Gilliland & Humphreys, 1943). A curious example is a paper by Elkine (1928). From his reproduction experiment he reports first an average deviation, then the means of over- and underestimations, respectively, for the same standard, but does not indicate the number of subjects each is based upon.

This flaw in data reporting appears to be based on the supposition that subjects „should“ react veridically, but are committing errors. The idea of a psychophysical function relating subjective (experienced, sensed) duration to clock duration, seems to have been alien to these researchers. However, the research mentioned is comparatively old. Contemporary research data are reported more completely, though more recent inscrutable data can be found occasionally (Krishna & Saxena 1984; Strang et al., 1973). I am only grieved that perfectly good data do not lend themselves to a more detailed treatment.

Under- and overestimation

The expressions „under-“ and „overestimation“ are frequently encountered in the literature on time perception (some recent examples: Burt & Kemp, 1994; Petrovici & Scheider, 1994; St. Jean et al., 1994), however the terms are ambiguous. The ambiguity is clearest in reproduction experiments, for example, does „overestimation“ refer to the standard, thereby implying that the reproduction is comparatively long, or to the reproduction, implying that it is comparatively short? The same problem appears in the other psychophysical methods. In estimation, for instance, an estimate exceeding the standard (e.g., 5 clock seconds are estimated as 6 seconds) may be due either to the time interval being experienced as longer, or to the unit of measurement (subjective seconds) being too short.

Overestimation, for instance, usually serves to indicate that an estimate, a production, or a reproduction, exceeds the standard. [For a thorough analysis, see, e.g., Bindra & Waksberg (1956) or Grant (1967a, b).] The problem becomes clear when a duration is experienced as longer than a standard. One would say that it is overestimated when using an estimation method (5 clock seconds are estimated as 6 seconds), and underestimated when using production (5 clock seconds are produced when 6 - subjective - seconds are requested), a rather unsatisfactory description of what is going on. Accordingly, I would recommend that the terms „underestimation“ and „overestimation“ be avoided in favor of a precise description: a 10-s interval is verbally estimated as 12 s; when required to produce a 8-s interval the subject produced 7 s; 5 s are reproduced as 4.5 s, etc.

Data presented as ratio of response over standard

Presenting data as ratios of responses over standard durations, for instance, reproductions to standards, runs counter to the concept of a psychophysical function. Such ratios obscure the course of growth of subjective duration with clock duration.

However, ratios of this kind may prove useful in some contexts. For instance, in certain group comparisons, as in Block & Zakay's diagram on gender differences (see Block & Zakay, this volume, chapter 9, Figure 4). Also, from a statistical point of view there is an advantage, homoscedasticity, counteracting Weber's law. (A better way of achieving the same thing is to take logarithms.)

Design of experiments

Two common weaknesses in experimental design on time perception are brought up here: too few standards, and too small a range of standards. Their consequences are similar.

No function

To be able to construct the psychophysical function at least four values (points) are necessary. With fewer points a possible curvilinearity cannot be discerned. If only one point (one standard) is used, for instance one second (Lhamon & Goldstone, 1956), no general conclusion can be drawn.

The time-order error

As early as 1868 Vierordt discovered that deviations from the standard are not unidirectional: reproductions of short durations are longer than the standard, reproductions of long durations shorter (Vierordt's law).

The observation that identical stimuli presented in succession are not experienced as equal is called the time-order error (see, e.g., Hellström, 1985) and is not confined to the continuum of time, though its roots may be different for time and other continua. The time-order error demonstrates clearly that sensations most often are not veridical, and that their deviations from physical magnitudes are systematic rather than haphazard.

Because reproductions of short intervals are longer than the standard, and of long intervals shorter, there is one point (or region) where subjective and physical magnitudes agree: the indifference interval (Woodrow, 1951). It follows that a researcher may risk concluding erroneously that time perception is veridical if he or she uses only one standard and this happens to lie in the indifference interval. But even in more extensive experiments the indifference interval is not a useful concept, because it depends on a number of factors, e.g., the range of standards, and its location is thus tied to the particular experimental design.

Lack of generality

Sometimes only a small range of standards is used. Wearden (1991), for instance, in order to avoid chronometric counting, worked only with durations less than 1 s. It is doubtful whether findings restricted to such short durations will hold up for longer durations as well. Findings based on too few standards or a too restricted experimental range do not allow generalizations.

Assumption of a linear psychophysical function

The psychophysical function as such is assumed linear

A linear psychophysical function is a special case of Stevens' power law $\Psi = \kappa(\Phi - \Phi_0)^\beta$, where Ψ denotes subjective and Φ physical (clock) duration with the exponent $\beta = 1$. (κ is a scale constant and Φ_0, the subjective zero, states the point in real time to which the subject locates the start of the duration.) In a survey of more than 100 papers about 500 exponents were listed with an average value of .9, rather than unity (H. Eisler, 1976). Not all researchers share my opinion that subjective time is not a linear function of clock time (see, e.g., Allan, 1978). Of course, .9 is not far from 1; sometimes the experimental and computational precision may be insufficient to distinguish between these values.

It is interesting to note that even researchers who are aware of subjects' lack of veridicality often assume a linear function, though with a slope that differs from unity. Varying rates of subjective time would correspond to varying values of the scale constant κ, with β fixed at one. It seems that the curvilinearity of the psychophysical function for time is difficult to accept. Among other reasons, this may be because a biological „clock" is assumed, and any clock is supposed to tick in proportion to physical clocks. For an exponent less than unity, however, the function is decelerating, implying a decreasing rate of „ticks."

Linearity and ratio-setting

In a ratio-setting experiment the values set by the subjects are almost always linear functions of the standards. (This is a general finding, not limited to time perception studies.) Many researchers assume as a consequence that the psychophysical function must be linear. But this conclusion is false. It may derive from a mistaken identification of the adjusted physical values and subjective magnitudes. As I was able to prove (H. Eisler, 1974), the above-mentioned linearity entails a power function whose exponent is determined by the slope of the straight line. As will be shown later, this is of particular importance for duration reproduction; subjective duration may be considered a variable intervening between standard duration and reproduction (H. Eisler, 1987).

The stimulus error

The stimulus error is committed when a stimulus is judged according to its physical rather than its experienced magnitude. The judgment of a percept as such is corrupted by some (usually learned) knowledge of the stimulus (see, e.g., Boring, 1950). A striking example is an experiment carried out by Teghtsoonian (1965). She had her subjects judge circle areas either as „proportional to its area in terms of square units" or „proportional to its size as it appears to you." Whereas the judgments according to the first instruction resulted in a linear function, a curvilinear power function with an exponent of .76 was obtained for the second. Thus the subjects could be instructed to commit or not commit the stimulus error. Assuming that perception as such works independently of instruction (which I do here for the sake of argument), the immediate sensation can be „corrected" into agreement with learned magnitudes. I would like to give a more personal illustration. Despite having lived in the same house for almost thirty years, every time I take the elevator from the basement parking area to the fourth floor (were my apartment is) I have the feeling of having reached my destination already between the second and third floor and am surprised that there is still some way to go I seem never to learn. But there is hardly any doubt that I could learn the objective duration if I really wanted to do so-thereby committing the stimulus error. Michon's (1978) distinction between analytic and impressionistic estimations offers a close parallel, though the basic reasoning differs. Clock durations are well learned and one has to be very careful when instructing subjects in order to get at their direct, uncorrupted percepts.

In scalar timing experiments (Gibbon et al., 1984) physical durations are taught by reinforcement. In my opinion, these experiments do not tap time experience directly; the animals commit the stimulus error. In this connection the drug effects illustrated by Block & Zakay (this volume, chapter 9) are very interesting. The mean response duration for an earlier learned standard is changed. One may ask whether this change is due to an altered time experience or whether the drug affects the capability to commit the stimulus error.

Bisection of temporal intervals

Attempts to determine the psychophysical function for duration have been made by having animals (or humans) bisect time intervals (Allan & Gibbon, 1991; Church & Deluty, 1977). In a bisection experiment the task is to find a duration in the middle of two presented durations, a short and a long. However, the midpoint can be interpreted in a number of ways: the same distance to the endpoints linearly, logarithmically, or as the same similarity, etc. Thus the conclusion to be drawn from bisection experiments depends on arbitrary (implicit) assumptions (H. Eisler, 1980).

The parallel-clock model

Choice of method

The experimental method my research group is using almost exclusively is duration reproduction. In a typical experimental trial, the subject is presented with one of a number of standard durations, indicated by noise. After a short pause (a silent interval), experienced as interruption, the same noise resumes and is shut off by the subject pressing a button when she/he has experienced that the second duration (the reproduction, that is, the noise after the pause) has lasted as long as the standard. Every standard is presented six times in pseudo-random order, and reproduced. Typically there are 10 standard durations ranging from 1.3 to 20 s in logarithmic steps.

The reasons for using the method of reproduction are threefold: (a) I wanted to carry out psychophysical scaling experiments with animals (which supposedly cannot handle numbers), (b) this method exhibits less scatter than other methods (Zakay, 1990), and (c) it probably comes closest to pure biological time perception, because it is number-free. Thereby it paves the way for animal experiments and avoids possible distortions caused by the subjects' number behavior. A more general description of how I got interested in time perception and duration reproduction can be found in H. Eisler (1990b).

The model

The common assumption regarding a duration reproduction task is that the standard duration is stored in memory, and the reproduction is continuously compared with this memory. The parallel-clock model disposes of any memory. Instead, two sensory registers are assumed; one accumulates subjective duration units from the beginning of the standard, the other from the beginning of the reproduction. Both accumulations are terminated simultaneously by the button press when the subject experiences that the reproduction has lasted as long as the standard. This occurs when the difference between the contents of the two registers equals the contents of the second register (H. Eisler, 1975; H. Eisler & A.D. Eisler, 1991). Figure 1 (p. 71) illustrates the model.

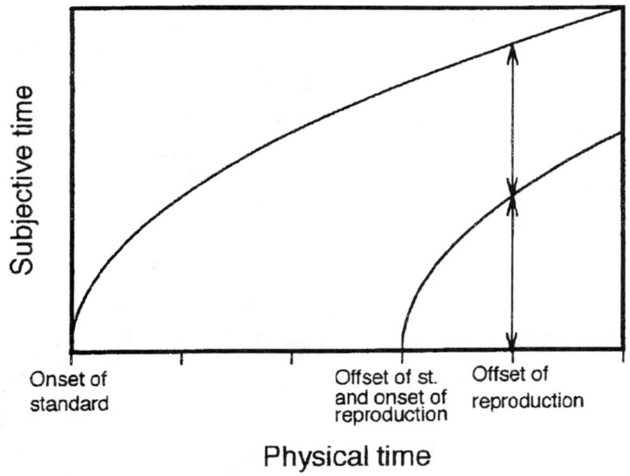

Figure 1: Duration reproduction according to the parallel-clock model. Subjective versus total physical duration (left curve) and versus reproduction duration (right curve). When the difference between these two subjective durations (upper arrow) equals the subjective reproduction duration (lower arrow), the subject reports equality between standard and reproduction by shutting off the sound (H. Eisler, 1989, p. 13)

The model is very powerful inasmuch as it allows the determination of the parameters of the power function from the number-free reproduction data. What is reproduction (equal-setting) from the point of view of the subject is halving as seen by the researcher. As is clear from Figure 1, the subjective counterpart of the reproduction is half of the subjective *total* (standard + reproduction) duration.

I have already mentioned the linearity of plots of ratio-settings against standards. A straight line implies a constant exponent over the whole experimental range. In applying the parallel-clock model, reproductions have to be plotted against *total* duration, i. e., the sum of standard and reproduction. Such a plot, containing real data, is shown in Figure 2 (p. 74). It consists of three rather than one straight line. The two breaks in the raw data imply a single break in the psychophysical function which can be described mathematically as a change, at a certain point in time, of the scale constant κ or the subjective zero Φ_0 or both[1]. The location of the break varies greatly between subjects.

Of all data sets studied, a few do not display a break. Interestingly, the latter also show an exponent close to one. I have a suspicion, though no clear evidence, that the subjects who produced these data were using some premeditated strategy, like counting. Otherwise, both data, the subjects' general behavior, and the subjective impression when I acted as a subject, speak strongly against the assumption that counting is used in duration reproduction experiments.

Computational details can be found in H. Eisler (1975) and later, more sophisticated methods in, e.g., H. Eisler & A.D. Eisler (1992). Three criteria confirm the model: 1. the parallelism of the two outer lines testifies to the constancy of the exponent β over the experimental range, 2. so does the linearity of the middle line, and 3. in the raw data plot (reproduction versus total duration) the value of the abscissa at the lower break has to agree with the value at the ordinate of the upper, which is the point in time of the single break in the psychophysical function.

The effect of a sudden change in the scale constant κ is described for fictitious data (to achieve more clarity) in Figure 3 (p. 75). Unfortunately, the method of reproduction does not allow a determination of the

---

[1] It may be worth mentioning that breaks are not only found when using reproduction; they appear also in magnitude estimation experiments (see, e.g. Figure 5 in H. Eisler, 1975). Psychologically, the break may mean a sudden shift in the subjects frame of reference; neurophysiologically, a change from one neural loop to another (Thatcher, 1979; Thatcher & John, 1977). But here I am transgressing the limits of psychophysics.

absolute value of the scale constants. A possible difference in the basic rate of accumulation of time units cannot be discovered, as long as the decrease of the rate (the exponent β is the same. However, the ratio between the scale constants of the upper over the lower segment of the psychophysical function can be determined. In the following this ratio will be denoted by α.

I wish to point out that the break is not a prerequisite for the validity of the parallel-clock model, though it can be regarded as a confirmation. What the model does is to describe in a parsimonious way the general finding that standards are not reproduced veridically.

Applications of the parallel-clock model

Different tasks

Absolute judgments of single stimuli

Results from absolute judgments („long" or „short") of single stimuli can be explained in detail by assuming that the subject in reality is considering two successive durations. At the nth trial one sensory register contains the subjective duration corresponding to the sum of the durations presented at trials n-1 and n, and the other register contains the subjective duration presented at trial n. After the response is given, the first register is cleared and accumulates, during trial n+1, the duration presented then. At the same time, this duration is accumulated on top of the previous one in the second register. Thus, the registers contain alternately the total and the presented duration (H. Eisler, 1981b).

Duration discrimination

Allan (1977) asked her subjects to judge which of two sequentially presented visual stimuli, taken from a set of four, was the longer. The parallel-clock model, together with Thurstone's model for pair comparisons (see, e.g., Guilford, 1954), accounted best for the data among a number of investigated models (H. Eisler, 1981a).

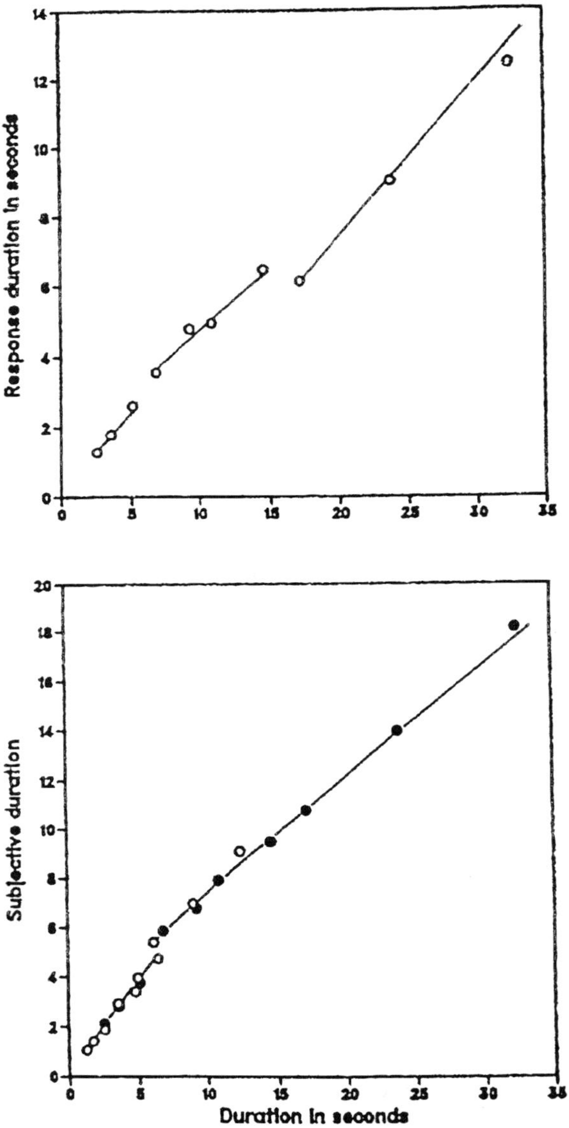

Figure 2: Upper panel: reproduced versus total duration; Lower panel: the psychophysical function (subjective versus physical duration) for one subject. ● denote total durations and o reproductions in the lower panel. Note that the break at about 6 s (lower panel) corresponds to a break at the same duration at both the abscissa and the ordinate in the upper panel.

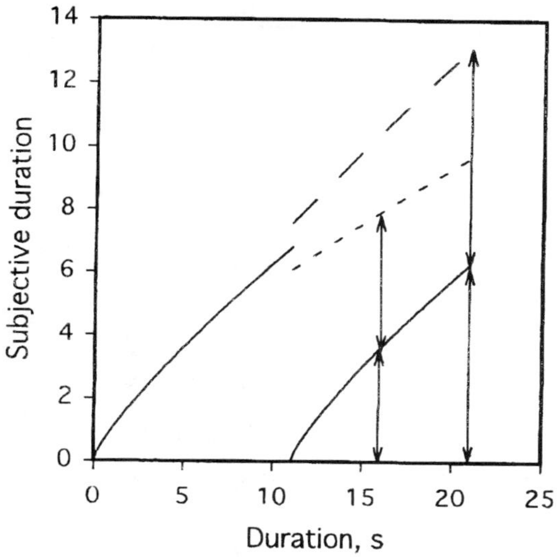

Figure 3: Illustration of duration reproduction for two fictitious subjects with different α values, reproducing a standard duration of 10 s with about 5.5 and 11 s, respectively.

Duration reproduction in H. M., an individual with memory deficit

The power of the parallel-clock model is well demonstrated with reproduction data (Richards, 1973) from an individual (known in the literature as H. M.) with an almost complete loss of long-term memory caused by a bilateral medial temporal-lobe resection (Scoville & Milner, 1957).[2] Figure 4 (p. 76) shows again two segments of the psychophysical function. But H.M.'s data differ from the usual in that there is no break at a fixed point in time; there is a large overlap between the two segments.

Differences in parameter values

One of the advantages of the parallel-clock model is that differences between groups or experimental conditions can be attributed to parameter differences in the psychophysical function. Variations in the exponent β for a large number of studies are given in H. Eisler (1976). More recent findings from experiments by my research group are described below.

---

[2] R.A. Block inspired me to subject these data, which I had treated cursorily before (H. Eisler, 1976), to a closer study.

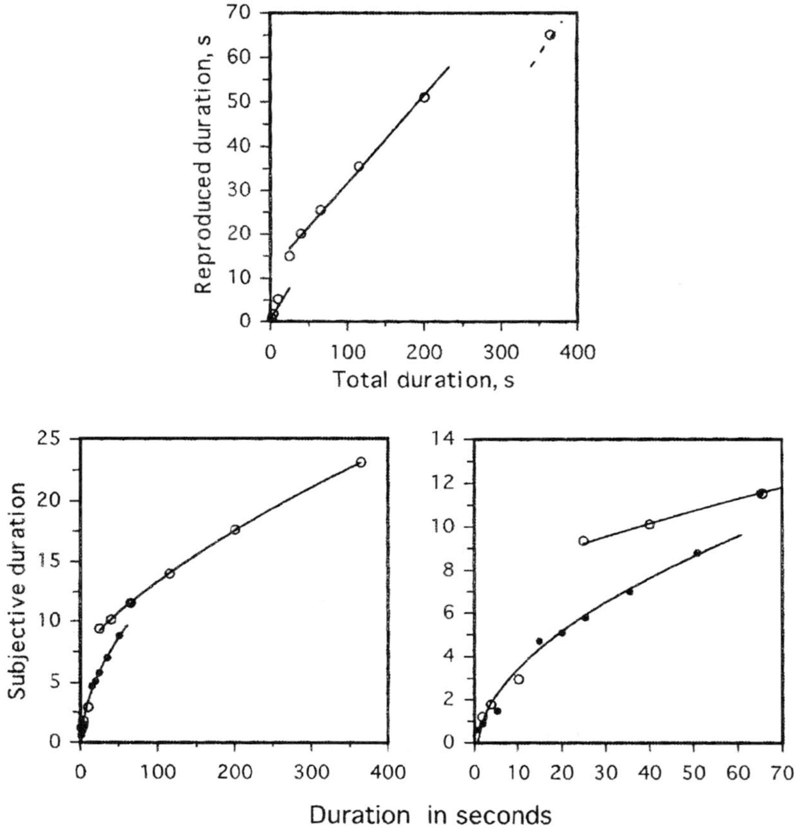

Figure 4: Data for H. M. (from Richards, 1973), processed according to the parallel-clock model. Upper panel: reproductions plotted versus total duration. Lower panels: the psychophysical function, where o denotes total and • reproduced duration. The right-hand panel is an enlarged part of the left-hand panel for shorter durations, demonstrating the overlap of the two segments.

Stimulus intensity

The intensity of the noise indicating the durations affects the reproductions; the louder the sound, the shorter the reproduction. This finding in the raw data can be attributed to the exponent; the greater the sound intensity, the smaller the b values (H. Eisler & A.D.Eisler, 1992). One might be led to believe that our empirical result is in agreement with Treisman et al.'s (1994) assertion, deduced from their time perception model which comprises a pacemaker and an arousal-sensitive calibration unit (CU): „as the intensity ... of sensory stimulation increases it will increase the arousal of the CU and so increase the calibration factor. Thus the louder the tone used to reproduce a time interval, the shorter the reproduction would be" (p. 250). But their argument - if I have understood it correctly - supposes that the standard is presented at a lower intensity, as opposed to our experiment. A very tentative explanation will be given in the section Neurophysiological speculations.

Gender

For durations below about 4 s on the average there seems to be no difference between male and female subjects. Longer durations, above the break, yield longer reproductions for female subjects. This difference can be attributed to a higher value of a, the ratio of the scale constant of the upper to that of the lower segment. There is no difference in the exponent (A.D. Eisler, in press; A.D. Eisler & H. Eisler, 1994; H. Eisler & A.D. Eisler, 1992).

Age

The findings for age parallel those for gender. The α value is higher for older subjects (A.D. Eisler & H. Eisler, 1994).

Accordingly, young men reproduce shortest and older women longest (and most veridically).

Differences between African immigrants and Swedes

A.D. Eisler (1992) showed that male African immigrants reproduce shorter than male native Swedes. The Africans had smaller values of the exponent. The finding could be generalized to both sexes (A.D. Eisler, in press). Whether this difference can be attributed to cultural differences, to the effects of emigration (e.g., stress caused by forced assimilation and loss of own culture, „detribalization"), or to other causes is a still unsolved problem (see also H. Eisler, 1995).

Personality: type A and type B behavior

One would expect that time perception varies between individuals with different personalities. People who display time urgency, which is assumed to be a component of the Type A behavior pattern, should differ from Type B individuals who do not show this behavior pattern. However, in our duration reproduction experiment (A.D. Eisler & H. Eisler, 1994) there was no difference between Type A and Type B personalities. Of course, it may be that the basic tempo, expressed in the unobtainable absolute values of the scale constants (cf. Treisman et al.'s pacemaker), differs between the two personality types[3]. Another psychophysical method may be more successful in catching possible tempo differences.

Time perception in animals

To my knowledge all studies of time perception in animals are based on discrimination, or on reinforcing responses to clock time. As I mentioned, one cannot get at the unadulterated time perception this way (see also Luce, 1984).

Beside the difficulty of communicating, one of the psychophysical scaling problems is an animal's inability to use numbers. But this was exactly why I chose the number-free method of reproduction. I succeeded in teaching rats (two out of eight) to reproduce time intervals between 1.3 and 20 s and obtained an exponent of the power function of about .5 (H. Eisler, 1984). An exponent of this size would be in agreement with Michon's (1978) „impressionistic" estimations. It is of interest to note that their data, like human data, exhibited a break[4].

There were three phases to the experiment. In the first phase the rats just learned that bar pressing led to reward. In the second phase, as in a typical reproduction experiment, there was a noise of varying duration, followed by a short pause, after which the sound resumed. The aim of this phase was to teach the rats to attend to the interruption between the first and the second duration. Their task was to suspend bar pressing until after the pause. A bar press during the first duration had two effects: the duration was lengthened by 4 s (to avoid superstitious chaining), and the reward was forfeited. Pressing the bar at any moment after the pause terminated the second duration, and, if there had been no bar press during the first duration, the rat was rewarded with sweetened milk. The third phase, the main experiment, was duration reproduction. The rats were rewarded when they reproduced the standard between a lower and an upper bound. The bounds were determined by a duration reproduction experiment with human subjects; for every standard, the shortest and the longest reproductions given in any trial by any of the subjects were taken as bounds.

---

3 The same goes for possible age and gender differences in tempo, assuming that the individual tempo is the same for standard and reproduction.

4 Complying with R. Levine's wish, I give a brief account of the experiment.

Some remaining problems

The effect of the inter-stimulus interval (ISI)

An experiment carried out recently (H. Eisler, Eisler, & Camara, 1994) for ISIs ranging from 40 to 600 ms showed drastic effects: 1. duration reproductions increase with the length of the ISI for all standards; 2. the increase exceeds ISI; 3. the increase for a fixed ISI is not constant, but larger the longer the standard duration; and 4. the increase goes to an asymptote. (The reproduction difference between 40 and 150 ms is much larger than that between 300 and 600 ms.) In these experiments only four ISIs were used. A repetition of the experiment is planned to obtain more points for the reproduction increase as a function of ISI. Also, longer ISIs will be employed to check whether the function really is asymptotical.

Number of standard durations and scatter

It is often difficult to determine the exact position of the break (H. Eisler, 1990a). In a bid for greater precision, the number of standards was increased from 10 to 19 in one series of experiments (A.D. Eisler & H. Eisler, 1994). This was in vain, however, because the scatter increased. The SDs of the raw data were larger and the fit of the parallel-clock model was worse. The cause of this deterioration has to be investigated; it may be due to the number of standards, or to the longer time required for the experiment.

Durations indicated by other modalities

In all work carried out by my research group on time perception, durations were indicated acoustically. There seems to be a difference between acoustically and visually indicated durations (H. Eisler, 1976; Goldstone & Goldfarb, 1964) which should be explored in more detail.

Diverse remarks

Paradigms

In time perception studies a distinction is made between prospective and retrospective paradigms. In a retrospective paradigm the subject does not know beforehand that she or he will have to judge a duration. Only after its termination, perhaps a long time after, will the pertinent instruction be given to the subject. In a prospective paradigm the subject is aware that she or he has to judge a forthcoming duration and is assumed to direct her or his attention to the flow of time. The latter paradigm is used in the majority of studies because after having given one retrospective judgment, the subject cannot be used further. Therefore the parallel-clock model cannot be applied properly, since it requires more than one empirical point from each subject. It can be used approximately for group data, as shown by H. Eisler (1976) for data collected by Pöppel (1972), whose subjects reproduced only one standard, and only once. But the large interindividual differences make the determination of a precise psychophysical function impossible. Remember that the position of the break varies greatly between individuals.

Durations filled with regular intermittent stimuli

When durations are indicated by a series of regular intermittent stimuli (e.g. clicks), the exponent is close to one (H. Eisler, 1977). I interpret this finding as a constant resetting of the biological clock, counteracting the deceleration that takes place when a duration is indicated by a continuous stimulus.

Miall's model

I would like to direct the attention of time perception researchers to Miall's (1989) model for duration memory, which is based on „beats." A time interval is „recorded" between the points in time when the same group of neural oscillators, with slightly different frequencies, are in phase at the start and at the end of a time interval. When this occurs again for the same group of oscillators, the new time interval equals the previous. Miall (1992) has recently generalized his model. Lebedev (1990; 1991) has similar ideas.

Neurophysiological speculations

I am most interested in the course of the psychophysical function and its parameters. It seems useful to me to be able to determine the effect of experimental manipulations on these parameters, but I am somewhat reluctant to explain findings of this kind by means of cognitive or attentional models. Not that I find such models uninteresting, but they are beyond my field proper. That notwithstanding, I cannot refrain from certain neurophysiological speculations.

According to my view, the biological clock does not consist of oscillators with constant frequency in clock time. It consists of neural loops (H. Eisler; Thatcher, 1979; Thatcher & John, 1977) that require an increasingly longer time for each passage ($\beta < 1$), perhaps due to an increase in the refractory period of the neurons. A counter registers the number of passages (pulses). A smaller β value would imply that the process of a decreasing number of pulses per clock time unit occurs faster. Perhaps heightened arousal quickens this process by some kind of fatigue, thereby explaining the effect of louder stimulation (section „Stimulus Intensity“) and the differences between Africans and Swedes (see section above with this heading).

The breaks indicate a switch from one neural loop to another, thereby achieving the highest resolution for a given duration. I am inclined to interpret the large variation in the position of the breaks as indicating that there is no „master“ biological clock, but quite a number of „clocks.“ This idea is congruent with what John Michon mentioned. When confronted with the duration reproduction task, a subject haphazardly „grabs“ the neural loop that comes first.

Summary

The chapter first enumerates a number of „misunderstandings and bad habits“ among researchers dealing with time perception which I came across in the literature. They probably stem from a certain psychophysical naïveté. I then explain my interest in the psychophysical method of reproduction and present the parallel-clock model, which is based on this method. The model allows the determination of the parameters of the psychophysical power function. Thereafter the potential of this approach to time perception is demonstrated by referring to applications of the model. For instance, which parameters vary with age and gender. Finally, unsolved problems in time perception which invite a psychophysical approach are considered. The paper concludes with some general reflections, among others it considers the distinction between prospective and retrospective paradigms, and neurophysiological speculations.

References

Allan, L.G. (1977). The time-order error in judgments of duration. *Canadian Journal of Psychology*, *31*, 24-31.

Allan, L.G. (1978). Comments on current ratio-setting models for time perception. *Perception and Psychophysics*, *24*, 444-450.

Allan, L.G. & Gibbon, J. (1991). Human bisection at the geometric mean. *Learning and Motivation*, *22*, 39-58.

Bindra, D. & Waksberg, H. (1956). Methods and terminology in studies of time estimation. *Psychological Bulletin*, *53*, 155-159.

Boring, E.G. (1950). *A history of experimental psychology*. New York: Appleton-Century-Crofts.

Burt, C.D.B., & Kemp, S. (1994). Construction of activity duration and time management potential. *Applied Cognitive Psychology*, *8*, 155-168.

Church, R.M., & Deluty, M.Z. (1977). Bisection of temporal intervals. Journal of Experimental Psychology: Animal Behavior Processes, 3, 216-228.

Eisler, A.D. (1992). Time perception: reproduction of duration by two cultural groups. In S. Iwawaki, Y. Kashima, & K. Leung (Eds.), *Innovations in cross-cultural psychology* (pp. 304-310). Amsterdam, Netherlands: Swets & Zeitlinger.

Eisler, A.D. (in press). Cross-cultural differences in time perception: comparison of African immigrants and native Swedes. In G. Neely (Ed.), *Perception and psychophysics in theory and application*. Stockholm, Sweden: HSFR.

Eisler, A.D. & Eisler, H. (1994). Subjective time scaling: influence of age, gender, and type A and type B behavior. *Chronobiologia*, *21*, 185-200.

Eisler, H. (1974). The derivation of Stevens' psychophysical power law. In H.R. Moskowitz, B. Scharf, & J.C. Stevens (Eds.), *Sensation and measurement*. Dordrecht, Holland: Reidel.

Eisler, H. (1975). Subjective duration and psychophysics. *Psychological Review*, 82, 429-450.

Eisler, H. (1976). Experiments on subjective duration 1868-1975: a collection of power function exponents. *Psychological Bulletin*, *83*, 1154-1171.

Eisler, H. (1977). Attention to visually and auditorily presented durations. In S. Dornic (Ed.), *Attention and performance VI* (pp. 247-259). Hillsdale, NJ: Lawrence Erlbaum Associates.

Eisler, H. (1980). Psychophysical similarities between rats and humans. *Bulletin of the Psychonomic Society*, *16*, 125-127.

Eisler, H. (1981a). Applicability of the parallel-clock model to duration discrimination. *Perception and Psychophysics*, *29*, 225-233.

Eisler, H. (1981b). The parallel-clock model: Replies to critics and criticisms. *Perception and Psychophysics*, *29*, 516-520.

Eisler, H. (1984). Subjective duration in rats: The psychophysical function. *Annals of the New York Academy of Sciences*, *423*, 43-51.

Eisler, H. (1987). Subjective duration as an intervening variable. In M. Teghtsoonian & R. Teghtsoonian (Eds.), *Fechner Day 87* (pp. 77-81). Northampton, MA: International Society for Psychophysics.

Eisler, H. (1990a). Breaks in the psychophysical function for duration. In H.-G. Geissler, M.H. Müller, & W. Prinz (Eds.), *Psychophysical explorations of mental structures* (pp. 242-252). Toronto: Hogrefe & Huber.

Eisler, H. (1990b). Serendipity in animal experimentation: Examples from duration scaling in rats. *International Journal of Comparative Psychology, 3*, 137-149.

Eisler, H. (1995). The psychophysical function for time perception: Interpreting their parameters. In R.D. Luce, M. D'Zmura, D. Hoffman, G.J. Iverson, & A.K. Romney (Eds.), *Geometric representations of perceptual phenomena. Papers in honor of Tarow Indow on his 70th birthday* (pp. 253-265). Mahwah, NJ: Erlbaum.

Eisler, H. & Eisler, A.D. (1991). A mathematical model for time perception with experimentally obtained subjective time scales for humans and rats. *Chronobiologia*, 18, 79-88.

Eisler, H. & Eisler, A.D. (1992). Time perception: Effects of sex and sound intensity on scales of subjective duration. *Scandinavian Journal of Psychology, 33*, 339-358.

Eisler, H., Eisler, A.D., & Camara, G. (1994).The effect of the interstimulus interval on duration reproductions. In L. M. Ward (Ed.), *Fechner day 94* (pp. 30-35). Vancouver, Canada: International Society for Psychophysics.

Elkine, D. G. (1928). De l'orientation de l'enfant d'age scolaire dans les relations temporelles [Orientation in temporal relations of the child of school age]. *Journal de Psychologie Normale et Pathologique, 25*, 425-429.

Gibbon, J., Church, R.M., & Meck, W.H. (1984). Scalar timing in memory. *Annals of the New York Academy of Sciences, 423*, 52-77.

Gilliland, A.R., & Humphreys, D.W. (1943). Age, sex, method, and interval as variables in time estimation. *Journal of Genetic Psychology, 62*, 123-130.

Goldstone, S., & Goldfarb, J.L. (1964). Direct comparison of auditory and visual durations. *Journal of Experimental Psychology, 67*, 483-485.

Grant, E. (1967a). Problems of methodology in time estimation studies. *Alberta Psychologist*, *8*, 12-25.

Grant, E. (1967b). Problems of terminology and psychophysics in time estimation studies. Alberta Psychologist, 8, 18-37.

Guilford, J.P. (1954). Psychometric methods. New York: McGraw-Hill.

Hellström, A. (1985). The time-order error and its relatives: Mirrors of cognitive processes in comparing. *Psychological Bulletin*, *97*, 35-61.

Krishna, L. & Saxena, N.K. (1984). Perceived time: its relationship with locus of control, filled versus unfilled time intervals, and perceiver's sex. *Journal of General Psychology*, *110*, 275-281.

Lebedev, A.N. (1990). Cyclical neural codes of human memory and some quantitative regularities in experimental psychology. In H.-G. Geissler, M.H. Müller, & W.Prinz (Eds.), *Psychophysical explorations of mental structures* (pp. 303-310). Toronto: Hogrefe & Huber.

Lebedev, A.N. (1991). Perception and memory: the framework of psychophysiological theory. In A. V. Holden, & V. I. Kryukov (Eds.), *Neurocomputers and attention: Vol. 1. Neurobiology, synchronisation and chaos* (pp. 45-59). Manchester, UK: Manchester University Press.

Lhamon, W.T. & Goldstone, S. (1956). The time sense: Estimation of one second durations by schizophrenic patients. *A. M. A. Archives of Neurology and Psychiatry*, 76, 625-629.

Luce, R.D. (1984). Time perception: discussion paper. *Annals of the New York Academy of Sciences*, *423*, 78-81.

Miall, C. (1989). The storage of time intervals using oscillating neurons. *Neural Computation*, *1*, 359-371.

Miall, C. (1992). Oscillators, predictions and time. In F. Macar, V. Pouthas, & W.J. Friedman (Eds.), *Time, action and cognition: towards bridging the gap* (pp. 215-227). Dordrecht, Netherlands: Kluwer Academic Publishers.

Michon, J.A. (1978). The perception and evaluation of time from a psychological viewpoint. *Epistemologia*, *1*, 223-250.

Petrovici, J.-N., & Scheider, G. (1994). Das Zeiterleben bei hirnorganisch Gesunden und fokal Hirngeschädigten [Time perception in organically healthy and brain damaged subjects]. *Fortschritte der Neurologie und Psychiatrie*, *62*, 256-267.

Pumpian-Mindlin, E. (1935). Über die Bestimmungmm der bewussten Zeitschätzung bei Normalen und dementen Epileptikern [On the determination of conscious time estimation in normals and demented epileptics]. *Schweizer Archiv für Neurologie und Psychiatrie*, *36*, 291-305.

Pöppel, E. (1972). Oscillations as possible basis for time perception. In J. T. Fraser, F. C. Haber, & G. H. Müller (Eds.), *The study of time* (pp. 219-241). New York: Springer.

Richards, W. (1973). Time reproductions by H.M. *Acta Psychologica*, *37*, 279-282.

Scoville, W.B., & Milner, B. (1957). Loss of recent memory after bilateral hippocampal lesions. *Journal of Neurology, Neurosurgery, and Psychiatry*, *20*, 11-21.

St. Jean, R., McInnis, K., Campbell-Mayne, L., & Swainson, P. (1994). Hypnotic underestimation of time: the busy beaver hypothesis. *Journal of Abnormal Psychology*, *103*, 565-569.

Strang, H.R., Rust, J.O., & Garrison, G. (1973). Sex differences in short-term time estimation. *Perceptual and Motor Skills*, *36*, 1109-1110.

Teghtsoonian, M. (1965). The judgment of size. *American Journal of Psychology*,*78*, 392-402.

Thatcher, R.W. (1979). The neural representation of experience as time. In G.G. Haydn (Ed.), *Experience forms: their cultural and individual place and function* (pp. 3-22). The Hague: Mouton.

Thatcher, R.W., & John, E.R. (1977). The neural representation of time. In *Foundations of cognitive processes: Vol. 1. Functional Neuroscience* (Eds.), (pp. 165-179). Hillsdale, NJ: Erlbaum.

Treisman, M., Cook, N., Naish, P.L.N., & MacCrone, J.K. (1994). The internal clock: electroencephalographic evidence for oscillatory processes underlying time perception. *Quarterly Journal of Experimental Psychology: Human Experimental Psychology*, *47A*, 241-289.

Vierordt, K. (1868). *Der Zeitsinn nach Versuchen* [Empirical studies of time experience]. Tübingen, Germany: H. Laupp.

Wearden, J.H. (1991). Human performance on an analogue of an interval bisection task. *Quarterly Journal of Experimental Psychology: Comparative and Physiological Psychology, 43B*, 59-81.

Woodrow, H. (1951). Time perception. In S.S. Stevens (Ed.), *Handbook of experimental psychology* (pp. 1224-1236). New York: Wiley.

Zakay, D. (1990). The evasive art of subjective time measurement: some methodological dilemmas. In R.A. Block (Ed.), *Cognitive models of psychological time* (pp. 59-84). Hillsdale, NJ: Lawrence Erlbaum Associates.

Chapter 5: The representation of change

JOHN A. MICHON

Introduction

The central tenet of contemporary time psychology is the following: in order to *stay tuned* to the course of events surrounding it and of which it is part, an organism must constantly exchange information with its environment. Even the most primitive species can cope with variable circumstances, but the real breakthrough in the course of evolution has been the emergence of a dedicated central nervous system, a 'programmable memory' that allows each individual of a species to learn and benefit by its own private experience. It provides independence with respect to both periodic and singular temporal events in ways that are consistent with those of the species to which the individual belongs.

In this light the experience of time is an advanced abstraction, an *ensemble of mental representations*, even if it ultimately derives from the basic and very concrete functional requirement to stay tuned to a dynamic, unfolding world. Mental representations of time enable us to achieve behavioral and cognitive coherence of successions of real events. The psychology of time seeks to understand, firstly, how a mind/brain is able to cope with the temporal contingencies - partly of its own design - of a dynamic world. Secondly, how a mind/brain can produce the rich phenomenology of time of which humans are consciously aware and which includes, among other things, the awareness of present, past and future, and of the rate at which time is perceived to pass.

The human experience of time appears to be vastly more complex than we would expect on the basis of the simple structure of physical spacetime. It is not surprising, therefore, that philosophers have hotly debated the ontological and epistemological status of time. As a somewhat distant witness to this battle I am left with a strong feeling that the flow of time may indeed be taken for an inherent property of reality, but that to *recognize* this flow for what it is, a conscious observer is required, tuned by evolution to the dynamic contingencies of the 'real' world. Assuming that the temporal structure of events is indeed physically determined, as a psychologist I wish to establish how this 'real' order is perceived and represented, what mechanisms are involved in the process, and how these succeed in generating the variety of phenomena of which we are normally aware.

Time figures prominently in everyday experience often in quite dramatic ways, the feeling of déja-vu, for instance, or the 'cliff hanger' phenomenon described by people who have miraculously escaped a critical and potentially deadly event. Such people frequently report having seen 'their entire life pass before their eyes' in a matter of seconds. As another example, some outstanding novelists, Franz Kafka, Marcel Proust, Thomas Mann and John Fowles among them, have succeeded in establishing narrative structures that leave their characters thoroughly disoriented in time. That these distorted temporal worlds - timescapes - are unconditionally convincing to us, the readers, suggests that we can adapt to quite extreme departures from the familiar clock time. What ultimately may be concluded from this is that time as experienced cannot be a unified concept deriving directly from something as simple as event order. Instead, we must conclude with Gorman & Wessman (1977) that

> "we should view temporal symbols not as distorted or altered perceptions of a determinate thing called 'time' but rather as representations chosen and constructed by individuals as apt expressions of their own life situations and feelings" (p. 238).

Time as information

Temporal relations *qua talis* have causal properties: events that do not distinguish themselves except for their relative position in a sequence of events such as the beats in meter and rhythm may nevertheless exert strong differential effects on behavior. We may observe this, for instance, when people are playing music or performing a dance. The perception of rhythm can be demonstrated in infants already very soon after birth and it may even be present in the unborn baby. Thus, temporal relations apparently contain vital *information* and both humans and animals use this information to guide their behavior. We may therefore assume that temporal relations are directly represented in the mind/brain. This assumption implies that the general features of human information processing should also apply to temporal information processing as well. This position has originally been put forward by Michon (1970) in what has become known as the *equivalence postulate*.

To illustrate this postulate I refer to Miller & Johnson-Laird (1976). In their monograph *Language and Perception* they argued that words and sentences serve as prescriptions for tuning a person's perceptual mechanisms to certain input conditions. Thus, the word *red* supposedly activates one's readiness to pay attention to the red stimuli in one's field of view. But, while there are perception-based tuning mechanisms for such dimensions as color or shape there is, the authors argue, no evidence for such a perceptual basis of temporal stimuli. Instead the temporal relations expressed in the (English) language seem to reveal a cognitive representation based on some simple form of first order logic. More recently, however, evidence has been accumulated which suggests that there are, after all, a good number of elementary perceptual mechanisms that are tuned to temporal relations between events: facial expressions and coordinated movements of the human body, for instance, appear to be perceived and encoded directly (e.g., Boltz, 1992; Freyd, 1987, 1992; Jones & Boltz, 1989). Certain dynamic properties of shape are also perceived and encoded directly (Leyton, 1989; 1992; Shepard, 1984).

Although some of the information regarding temporal relations is received without cognitive intervention, much is not. In order to accommodate the latter kind of (indirectly encoded) information, cognitive psychology relies on three theoretical constructs: quasi-visual mental images for coding spatial relations between items, abstract propositions for the encoding of meaning, and temporal strings (Anderson, 1983, p. 45) for encoding the order of items in a sequence. As an example involving the latter mode of representation, consider the temporal organization required to repair a vacuum cleaner. Unless one already happens to be an experienced vacuum cleaner mechanic, one benefits from treating the disassembling task as a temporal problem, that is, by relying on an order scenario. The many parts are to be located on the workbench in the *temporal* order in which they were removed, and successful reassembling mostly depends on following the inverse order. The important question of where and how the part is to be re-installed can frequently be answered by considering the shape of the various units: a nut simply "asks" to be mounted on a bolt.

What are the characteristics of the temporal string representation? Anderson has argued that they encode order but not other aspects of time such as duration or the passing of time. This, however,

> "is not to say that we cannot perceive or remember interval properties of a time sequence, but that such properties are not directly encoded in the structure of the temporal string ... Such information can optionally be encoded as attributes of the ordered elements." (Anderson, 1983, p. 49).

Following this line of thought, the question we are facing next is what then constitutes the 'temporal attribute' encoded in these strings? The simplest assumption consistent with the idea of temporal strings is that the physically given order of events provides this attribute. If temporal information is indeed encoded in terms of ordered strings, as is suggested by Anderson, with other temporal attributes such as duration or temporal locus added if required, the question becomes how order is encoded in the subject's mind. There is indeed a condition for ordered events to be temporally encoded: if one event is perceived and subsequently retained in memory, the next event to be perceived will necessarily *have the earli*er event as part of its (remembered) context. Thus the earlier-later relation is adequately represented as a stored vs. actual relation. But, even if order would turn out to be a necessary temporal stimulus, it cannot be the only relevant one. In the first place there is no valid reason to assume that the order of impressions will determine the impression of order. On the contrary, Dennett (1992) has argued forcefully that the temporal locus of an event is determined not on the basis of a single perceptual act, but on some averaging of a parallel stream of conflicting and continuously revised contents (Dennett & Kinsbourne, 1992, p. 183). At best these would provide partially ordered strings that would then have to be integrated in a single time line or narrative. In addition, abstract spatial or logical attributes may also provide cues that serve in those cases that order cues are not available or unlikely: if I put my cart in front of my horse, you will see my cart pass sooner than my horse.

In conclusion then, time can be but is not always directly encoded. And even if it is, it will usually be encoded in a more or less fragmentary fashion, episodes that do not necessarily provide a coherent history. Although there are primitive perceptual and motor functions that operate automatically in the presence of adequate stimulation, processing temporal information is indeed largely a deliberate cognitive activity, requiring a great deal of attention and strategic flexibility on the part of the individual. When and if temporal information is processed, order may be considered as the predominant stimulus attribute, but it may be an unreliable one. Certainly it is not the only one. Other, intrinsically non-temporal cues may be given a quasitemporal interpretation.

Formal representations of time

Being interested in how people construct their representations of a temporal world is not just a matter of intellectual curiosity. Equally important is the fact that the time has come that the human race must face the prospect of coping with other literate intelligences. These systems will evolve, and prudence demands that *we* define the terms of communication before *they* do. In short, it is necessary to think about how to go about making a time sense (Michon, 1989). The philosopher Nelson Goodman summarized the problem of establishing an appropriate representation of the world around us as follows: „People make visions and true visions make worlds“ (Goodman, 1984, p. 34). What he meant is that of all our attempts to construct representations of reality only a few are strong enough to support a coherent, instrumental view of the world. This is also true of our representations of time; some have proved so successful that we have become convinced that they simply must be telling us something essential about the temporal structure of nature. This conviction has made it possible to develop conceptual structures that are so coherent and consistent that we may call them theories of time (Michon, 1990). It is to these theories we must turn if we wish to formalize time so that we may eventually use it for our artificial sense of time.

The most conspicuous example would seem to be the clock metaphor. But the mechanical clock was initially less a time keeping instrument than a representation of the heavens, a planetarium allowing major social events to be coordinated. Its lack of precision did not allow the measurement and coordination of finely timed activities such as the singing of polyphonic hymns in the monasteries during the Middle Ages. As Szamosi (1986) pointed out in a particularly interesting monograph, the development of a timing system that allowed the division of brief intervals constitutes an artistic and chronometric achievement of the highest order. As such it is based on no less than five centuries of intense experimentation and daily practice. For an artificial time sense, however, the clock metaphor fails miserably: a robot that is driven by a clock is like the elaborate mechanical clocks and musical boxes that repeat themselves hour after hour.

Among the attempts at formalizing temporal relations, temporal logic takes a special place. In the 1960s temporal logics were developed by Prior and others, enabling us to determine the truth of propositions about complex relations between points in time or between time intervals (see e.g. Kamp, 1979; Van Benthem, 1983). Temporal logics - there are now several of them - find application as formal theories of action planning, an important problem area in artificial intelligence (e.g., Georgeff & Lansky, 1987; Shoham, 1988; Akyürek, 1992). Initially there were few ways of dealing with temporal relations between events. However, for many applications the absence of algorithms for temporal operations is not crucial. In many cases a computational process can be described, without loss of specificity, as an ordered list of static positions which succeed each other instantly. A game of chess can usually be described as a sequence of positions, except when players threatens to exceed their time limit.

Compare this with what is needed when driving an automobile approaching an intersection In that case computations about the consequences of different, partly simultaneous, serial actions and their outcomes is required (Aasman & Michon, 1992; Aasman, 1995). In this case the problem is to maintain the integrity or 'truth' of the knowledge base on which the action planning is based avoiding a computational explosion in which the number of possible connections between facts increases vastly more rapidly than the number of known facts itself. This seems to be one of the most fundamental problems that any plausible theory of intelligence that incorporates temporal phenomena will have to solve. Although the problem has been known in various disguises for many centuries, dating back at least to Zeno of Elea, it has recently attracted much attention in its formulation as the *frame problem* (e.g., McCarthy & Hayes, 1969; Pylyshyn, 1987). Some recent models for temporal planning seem to avoid this explosive problem altogether. They do not proceed from separate frames or actions with their conditions and results, but from string of elements, so-called *chronicles*, that represent an action sequence or episodic progression from its inception to the present instant (e.g., Lansky, 1987). Planning implies the choice of one of (perhaps infinitely) many possible worlds that are consistent with the course of events up to the present point in time, and that in addition meet the action constraints imposed by the present circumstances. This approach avoids the problem of dealing with the world as a succession of inert states that requires an explosive amount of truth maintenance if action is to be taken.

These are some of the trends toward formalized theories of time, planning and action. In many respects these trends seem to match the recent psychological insights into the mental representation of time. Thus they may well provide the foundations for a more explicit and detailed theory about the processing of temporal information and the way humans stay tuned to the dynamic world around them.

An illustration generic intelligent driver support

Let me now present you with a bird's eye view of a particular real time system called GIDS, which stands for Generic Intelligent Driver Support (Michon, 1993) GIDS is an appropriate context for my arguments because it is specifically designed as a real time support system. A driver support system is a system that will provide a vehicle operator with information in the form of warning, advice, instruction, or even active intervention. Their purpose is to make the driving task easier, safer, and less costly. GIDS was conceived and implemented as part of the EEC DRIVE Program, Phase I (1989-1992). Subsequent research has been carried out in DRIVE II under the name ARIADNE and still further development is foreseen in the context of DRIVE III, now in its planning phase.[1] The initial GIDS project resulted in recommendations for driver support systems in general and in two operational prototypes demonstrating the essential features of the GIDS concept, both under simulated and real world traffic conditions. A detailed description of the philosophy, development, and evaluation of the GIDS system can be found in the published version of the final, comprehensive report on this elaborate project (Michon, 1993).

Driver support systems derive their usefulness to a large extent from the fact that vehicle operators must cope with a growing amount of information of an increasingly complicated nature. This avalanche of information is likely to have an impact on almost every aspect of the driving task. It will affect route planning as well as navigation, maneuvering, or elementary vehicle control. Unless action is taken this information would eventually be presented to the driver in an essentially incoherent fashion, irrespective of meaning or urgency. An important function of a driver support system such as GIDS, helps to protect the driver from such uncoordinated, and under circumstances even contradictory, information. The important innovative feature of GIDS is that it is the first system ever to take into account, in an adaptive fashion, (some of) the intentions, capabilities, and limitations of the individual driver. For this purpose GIDS incorporates not only a knowledge base containing detailed scenarios for driving maneuvers such as overtaking, negotiating intersections, or merging into a traffic stream, but also a knowledge base containing information about the characteristics and behavioral history of individual drivers.

The driving task is conventionally divided into three major task levels: planning, maneuvering and control. The GIDS system incorporates support functions at each of these levels and, in addition, provides instructional support to novice drivers. A small but realistic set of driving (sub) tasks has been selected for in-depth study, including lane following, merging into a traffic stream, and overtaking. Driving takes place in a complex dynamic environment, the road system, under a set of fairly well-defined operating constraints. The GIDS project provides rec-

---

[1] Further details can be obtained from the Traffic Research Centre at the University of Groningen, P.O. Box 14, 9750 AB HAREN, The Netherlands.

ommendations for driver support in a wide variety of driving circumstances. The driver can interact with the GIDS system by means of a variety of displays and controls, including voice communication, a video-display and a special purpose keyboard. GIDS also includes two active, intelligent controls, the accelerator pedal and the steering wheel. Ideally GIDS should cover all of driver behavior under all conceivable circumstances. Since this is, at present, far beyond reasonable bounds, only a limited set of conditions is actually studied without, however, giving up the intended level of generality of the approach. This policy has resulted in the *Small World Paradigm* which restricts the number of situations to 43 and the number of elementary actions to which the GIDS system is responsive to 67 (from a set of altogether approximately 1700 distinguishable behaviors; see McKnight & Adams, 1970). The 67 actions are special in the sense that together they form a closed set: any combination of actions from the set will result in a situation that can be dealt with by at least one of the actions. In other words, in its small world GIDS will never be at a loss when it comes to deciding what to do next.

The functional architecture of GIDS includes in the first place a script-based *Analyst/Planner* which accepts information from dedicated input systems, both in-vehicle systems such as the speedometer, the car radio, the on-board navigation system, and road-side systems such automatic road signs and early warning signals. In the second place it incorporates a *Dialogue Controller* which regulates the communication between the driver and the GIDS system. The structure of the intelligence as embodied in the Analyst/Planner of the GIDS system can be described as follows. GIDS knows about the Small World, or more specifically, it has stored a representation of every Small World situation and of every type of event that can take place in that world. Each representation consist of a vector of variables (numerical and binary truth values) that unambiguously identify each event. These are actually the variables that GIDS will sample through its sensors, input/output devices and applications. Associated with each vector element is a range of acceptable values. As long as the driver's behavior remains within this range jointly for all variables, there should be no output and the driver will seemingly be operating in an open loop fashion. At all times, however, the driver may actively *request* information from the system, in which case the system will generally answer the request. As soon as one or more observed variables fall outside its acceptable range, GIDS will recognize this discrepancy. Every such critical situation is associated unambiguously with one or a series of messages. Here the term message is used in a generic sense: it may refer to a system-induced change in gas pedal response, a beep, a spoken phrase or a more or less complicated visual signal, a blinking light, a pictogram or a written message. Once the message has been passed on to the driver, the GIDS system will continue operating according to its overall way of operating: if the driver recovers from the critical situation or if the problem goes away as is. The system will stop delivering messages until the next critical situation emerges. Once a message has been determined, the problem is how it ought to be exchanged with the driver. In GIDS this problem is tackled by the Dialogue Controller. A first problem this system component must be able to face is the possibility that several critical events occur at the same time or so close in time that a *priority conflict* will arise. The second, related problem pertains to the *driver workload* created by the situation and by the message structure. And finally there is the problem of *allocation of messages* to the various interfaces of the GIDS system, in such a way that the driver is not confused or overloaded simply by the communication with the system as such.

The general issue is therefore one of *scheduling* or *planning*. Given an event or event sequence the performance of the driver depends on when and how to process the relevant information: how long does it take to read a road sign and how long to notice that a child is standing at the curb, waiting to cross? How long does it take to reach a decision to shift gears and how long to activate a brake? Many of these basic performance data are known, or else have been collected specifically for the project. What has not been studied, however, is how these elementary performance operations fit together into integrated maneuvers such as merging into a traffic stream or during an overtaking maneuver. In GIDS this problem is tackled by means of a method of determining the time and resource allocation for each separate action unit - turning right, changing lane, braking to a halt at a traffic light, or, for that matter, dialing a number on the car phone - in a particular maneuver. A similar analysis is made for each message that may be passed to the driver and for every message that the driver may pass to the GIDS system. These allocation patterns are stored and eventually retrieved together with the various situation and event representations.

Temporal issues in GIDS real time performance

Now, what sort of temporal issues do we face in our development of GIDS? First of all there is the operation of the Analyst/Planner component. As indicated before, this component is essentially a script- or case-based planner. It is based on a very detailed task analysis, and it embodies the more and less likely event histories that can be encountered in the context of a well-defined (but thus far quite limited) set of events that may arise in the 43 situations that may arise in the Small World. In principle GIDS should be able to cope with the Cartesian product of *situations* x *maneuvers* x *driver* actions but, currently it is much more restricted than that. Over the range of events for which it can provide the driver with support, the scripts in the Analyst/Planner have default values for certain inputs which the system derives from its sensors and its functional goal is to minimize, under all circumstances, the work load imposed on the driver when performing safely and effectively.

As an example consider the following description:

Function:	approach *intersection* to *turn right* [R at -3; slot -5/-1]
Function:	approach *pedestrian crossing* [present no message -4/0]
Conclude:	[R at -5]

In plain terms this just means the following: the system, and more specifically the Analyst/Planner detects that the driver it is supporting is approaching an intersection. On the basis of its topographical knowledge it knows that, in order for the driver to reach the destination she has specified prior to the trip, a right turn must be made at this particular intersection. GIDS knows - as is indicated in the bracketed expressions - that the supporting message for a right turn (R) must be delivered between 5 and 1 seconds before reaching a (standard or default) intersection, preferably at 3 seconds. Next the Analyst/Planner detects that it is approaching a pedestrian crossing located immediately before the intersection. This nature of the situation requires it to refrain from giving any message (with the exception of top priority warnings that overrule any other function) between -4 seconds and the instant at which the crossing is reached. The Analyst/Planner then concludes that the message for the right turn is to be issued 5 seconds before reaching the intersection. Thus a considerable part of the activities of the Analyst/Planner consists of organizing the temporal pattern of support messages in such a way that normal, humanly acceptable sequencing of task elements is guaranteed for the range of situations the GIDS system can cope with.

Once the Analyst/Planner has decided what messages to deliver and when is the time to do so, the Dialogue Controller takes over. Having to deliver a specific message in a specific situation, and given the current goal of the driver, the Dialogue Controller will estimate what (extra) workload is involved for the driver to understand this message and to take appropriate action. In other words, there is a certain time requirement for understanding the message and translating it into proper action. The Dialogue Controller takes care of this fine-tuning of the messages delivered by the GIDS system by means of a series of procedures. Its function is essentially to schedule the actions and messages delivered by the GIDS system in such a way that for every conceivable event or combination of events a non-conflicting interfacing with the driver is maintained. The operational basis for these procedures derives from the Model Human Processor (MHP) or Goals, Operators, Methods, and Selection (GOMS) concepts, which are crucial to much of Newell's work on task analysis and intelligent architectures (Card et al., 1983; Newell, 1990; 1992). These concept are closely related to classical ideas about time and motion studies on the one hand, and those of critical path analysis as it is applied in industrial management on the other. MHP and GOMS are assumed to model the information processing activity of humans. This assumption also lies at the root of the GIDS system: the control of the dialogue between the driver and the system exercised by the GIDS system is indeed based on an elementary version of the MHP. That is, the GIDS system operates on a realistic representation of its driver to the extent that the MHP is indeed a realistic model of human information processing. On the basis of explicit estimates about the time it takes to detect and discriminate the relevant stimuli, to perform a relevant cognitive operation, or to execute a relevant elementary motor act during the execution of a given task - such as driving - a critical path analysis is made. This means that an optimal trajectory is determined for sequencing the required operations so as to impose minimal information load on the operator. An interesting but taxing complication in this set-up is that the GIDS dialogue control requires scheduling for *uncertain* events in real time, meaning that there is no room for 'what if' strategies for finding optimal schedules. A simple, but effective, partial solution for this problem has been adopted: events and messages are also assigned a priority value which allows an urgent, potentially life-saving message to overrule schedules of lower priority order.

Conclusion

To claim that with the GIDS project we have succeeded in creating a time sense for artificial intelligences would be a gross exaggeration. We have demonstrated, though, that it is possible to impose constraints on an artificial intelligence which enable us to structure the system's internal representation of real time events and at the same time to take into account the way it processes the required information. These are *necessary* - although not *sufficient* - aspects of a time sense. Having implemented them in an operational driver support system proves that we have proceeded to the point where such a system can independently and dynamically interact with the external world.

References

Aasman, J. (1995). *Modeling driver behaviour in Soar* [Dissertation University of Groningen]. Leidschendam, The Netherlands: Royal PTT Nederland NV.

Aasman, J. & Michon, J.A. (1992). Multitasking in driving. In J.A. Michon & A. Akyürek (Eds.), *Soar: a cognitive architecture in perspective* (pp. 169-198). Dordrecht, The Netherlands: Kluwer.

Akyürek, A. (1992). On a computational model of human planning. In J.A. Michon & A. Akyürek (Eds.), *Soar: a cognitive architecture in perspective* (pp. 81-108). Dordrecht, The Netherlands: Kluwer.

Anderson, J.R. (1983). *The architecture of cognition*. Cambridge, MA: Harvard University Press.

Boltz, M. (1992). The incidental learning and remembering of event durations. In F. Macar, V. Pouthas, & W.J. Friedman (Eds.), *Time, action and cognition: towards bridging the gap* (pp. 153-163). Dordrecht, The Netherlands: Kluwer.

Card, S.K., Moran, T.P., & Newell, A. (1983). *The psychology of human-computer interaction*. Hillsdale, NJ: Erlbaum.

Dennett, D.C. (1992). *Consciousness explained*. Harmondsworth, Middlesex, England: Allen Lane/Penguin.

Dennett, D.C. & Kinsbourne, M. (1992). Time and the observer: the where and when of consciousness. *Behavioral and Brain Sciences, 15*, 183-247.

Freyd, J.J. (1987). Dynamic mental representations. *Psychological Review, 94, 427-438.*

Freyd, J.J. (1992). Dynamic representations guiding adaptive behavior. In F. Macar, V. Pouthas, & W.J. Friedman (Eds.), *Time, action and cognition: towards bridging the gap* (pp. 309-323). Dordrecht, The Netherlands: Kluwer.

Georgeff, M.P. & Lansky, A.L. (Eds.). (1987). *Reasoning about actions and plans*. Los Altos, CA: Morgan Kaufmann.

Goodman, N. (1984). *Of mind and other matters*. Cambridge, MA: Harvard University Press.

Gorman, B.S., & Wessman, A.E. (Eds.). (1977). *The personal experience of time*. New York: Plenum Press.

Jones, M.R., & Boltz, M. (1989). Dynamic attending and responses to time. *Psychological Review*, *96*, 459-491.

Kamp, H. (1979). Instants, events, and temporal discourse. In R. Baeuerle, E. Ugli, & A. Von Stechow (Eds.), *Semantics from different points of view* (pp. 376-417). Berlin: Springer.

Lansky, A.L. (1987). A representation of parallel activity based on events, structure, and causality. In M.P. Georgeff & A.L. Lansky (Eds.), *Reasoning about actions and plans* (pp. 123-159). Los Altos, CA: Morgan Kaufmann.

Leyton, M. (1989). Inferring causal history from shape. *Cognitive Science*, 13, 357-387.

Leyton, M. (1992). *Symmetry, causality, mind*. Cambridge, MA: MIT Press.

McCarthy, J.M., & Hayes, P.J. (1969). Some philosophical problems from the standpoint of artificial intelligence. In D. Michie & B. Meltzer (Eds.), *Machine Intelligence* (pp. 463-502). Edinburgh: Edinburgh University Press.

McKnight, A.J., & Adams, B.B. (1970). *Driver education task analysis. Volume I: Task descriptions* [Final Report, Contract Nr. FH 11-7336]. Alexandria, VA: Human Resources Research Organization.

Michon, J.A. (1970). Processing of temporal information and the cognitive theory of time experience. *Studium Generale*, *23*, 249-265.

Michon, J.A. (1989). Timing your mind and minding your time. In J.T. Fraser (Ed.), *The study of time: Vol. 6. Time and mind: interdisciplinary issues* (pp. 17-39). Madison, CT: International Universities Press.

Michon, J.A. (1990). Implicit and explicit representations of time. In R.A. Block (Ed.), *Cognitive models of psychological time* (pp. 37-58). Hillsdale, NJ: Erlbaum.

Michon, J.A. (Ed.). (1993). *Generic Intelligent Driver Support*. London: Taylor & Francis.

Miller, G.A. & Johnson-Laird, P.N. (1976). *Language and perception*. Cambridge: Cambridge University Press.

Newell, A. (1990). *Unified theories of cognition*. Cambridge, MA: Harvard University Press.

Newell, A. (1992). Unified theories of cognition and the role of Soar. In J.A. Michon, & A. Akyürek, *Soar: an intelligent architecture in perspective* (pp. 25-79). Dordrecht, The Netherlands: Kluwer.

Pylyshyn, Z.W. (Ed.). (1987). *The robot's dilemma: the frame problem in artificial intelligence*. Norwood, NJ: Ablex.

Shepard, R.N. (1984). Ecological constraints on internal representation: Resonant kinematics of perceiving, imagining, thinking, and dreaming. *Psychological Review*, *91*, 417-447.

Shoham, Y. (1988). *Reasoning about change*. Cambridge, MA: MIT Press.

Szamosi, G. (1986). *The twin dimensions: inventing time and space*. New York: McGraw-Hill.

Van Benthem, J.F.A.K. (1983). *The logic of time*. Dordrecht, The Netherlands: Reidel.

Part IV:
Cultural differences in psychological time

Chapter 6: Psychology of time from a cross-cultural perspective

HEDE HELFRICH

Introduction

All cultures have concepts of time. Physical phenomena and significant life events are temporally classified, and there exist mental representations of the flow and structure of time. Social life is submitted to temporal rules: work is organized in temporal patterns; certain times are reserved for festivals; rituals follow rules of duration, simultaneity and succession; speech, music, and dance show temporal patterns. There are, however, considerable differences in the „social representations" (to borrow a term from Durkheim, 1898, or Moscovici, 1981) of time and in the social treatment of time (cf. Thomas & Helfrich, 1993). Roughly, four dimensions can be identified to account for the cultural variations. I will call them „objective time," „subjective time," „social time" and „biological time." *Objective time* refers to the mental representations of the flow and structure of time, that means that „time" is made into a mental „object" by the mind of the thinking subject.[1] *Subjective time* refers to the individual time horizon or temporal perspective. *Social time* refers to the organization of time in social life. And *biological time* refers to biological life events such as birth and death, endogenous cycles, oscillations and rhythms within biological processes as well as to their entrainment by external pacers (cf. McGrath & Kelly, 1986, p. 42).

It might be stated, that each culture has developed preferred positions on each of the four dimensions, and the resulting set of value preferences determines how time will be construed within a specific culture (see also McGrath & Kelly, 1986, p. 18). In this, the four dimensions are assumed not to be orthogonal or mutually independent, rather, there

---

[1] It should be noted, that the term „objective time" is here, unlike in chapter 1, not used in the meaning of physical time.

might be typical patterns of preferred positions. For instance, a future-driven time horizon usually is contingent with a linear concept of time. However, the preferred positions are predominant positions which do not consistently hold within all life contexts.

Objective, subjective, social and biological time might be conceived as forming successive levels of accessibility to the conscious mind, with objective time most and biological time least accessible to conscious thinking. While the relationship between objective and subjective time seems to be obvious, the relationship between biological time and the remaining dimensions might not be unanimously accepted. On the one hand, one might favor a radical rupture between biological and cultural aspects of time conceiving them as unrelated, or on the other hand, one might adhere to an evolutionary view and look for a continuity from biological to objective time (cf. Richelle et al., 1985, p. 75, and Lejeune & Richelle, this volume, chapter 8).

Objective time

Mental representations of the flow and structure of time may vary across cultures with respect to three aspects. The flow of time may be conceptualized either as cyclic or as linear (cf. Nakamura, 1968), the time representation can be concrete or abstract, and time may be considered continuous or divisible.

Cyclic versus linear flow of time

The course of time can be represented symbolically either in the form of a circle, as a „cyclic" recurrence of uniform events, or in the form of an arrow, viewing time as a „linear" progress without repetition. The cyclic time conception focuses on rhythmic events such as day and night, or summer and winter, and is typical for archaic societies, in which human actions are assumed to be a mythical reconstruction of sacred actions (cf. Eliade, 1953), and for the religious systems of Hinduism and Buddhism (cf. Nakamura, 1968). The linear time conception historically traces back to the monotheistic systems such as Judaism, Christianity, and Islam, and is closely related to the Western concept of progress. Although in Western cultures the conception of a linear time generally prevails, there are settings which are permeated by cyclic conceptions, e.g., the feasts of the Christian ecclesiastical year.

Closely related but not identical to the question of linearity versus cyclicity is the question, whether time is regarded as „uni-directional" or as „bi-directional." At first glance, one might assume that the linear view of time implies that time is unidirectional as opposed to the cyclic view which is indifferent with respect to directionality. This is not correct in terms of classical physics: Newtonian physical science assumes bidirectionality as opposed to Einsteinian physics favoring unidirectionality. However, it seems that independently of the scientific debate, most people from Western cultures prefer the unidirectional concept.

Abstractness of time representation

Time may be divided or „measured“ in terms of observable events such as river inundation or moon cycles, where often the individual dates linked to different events are not mutually integrated. The opposite of this concrete representation of time is the abstract representation according to which time units - such as minutes, hours or weeks - are independent of environmental events and are compatible with one another (cf. Friedman, 1990). Although the abstract representation of time may be characteristic for Western cultures, it does not consistently hold: when people from Western cultures are asked to date past events, they relate them to other events, but often cannot indicate the temporal distance to the present time (cf. Friedman, 1990).

Structure of time

Another aspect of the concept of time has to do with whether time is to be regarded in a holistic view as continuous, i.e. indivisible, or in an atomistic view as discontinuous, i.e., divisible. Related to the holistic view, time is conceived as having homogeneous duration, whereas the atomistic view implies that time is regarded as successive, i.e., as a series of successive instants (cf. McGrath & Kelly, 1986, p. 22). The preferred position strongly affects the subjective experience of time. From a holistic viewpoint, the present overlaps with the future and the past, whereas, from an atomistic viewpoint the flow of time can be divided into single events with „empty“ temporal spaces between the „filled“ segments of time.

Subjective time

All human individuals have concepts with respect to their own time, that is the time associated with present activities and future goals. This subjective time representation, which relates the present to the future, is called „time horizon“ or „temporal perspective.“ Although it is subjective in referring to the life of the individual, time horizon or time perspective is an essential characteristic of culture (Jones, 1988, p. 22).

Time horizon bears a strong relationship to achievement orientation (Jones, 1988) which is generally defined in terms of goal-directed behavior. While behavior is always carried out in the present, goals typically refer to some point in the future, and goal-directed behavior refers to those forms of behavior which are expected to increase the probability of attaining the desired goals. Such behavior is called *instrumental*, as opposed to *expressive* behavior which, having its end in itself, bears no apparent relationship to any goal.

However, although goals are always localized in the future, the future, relative to the present, may be near or far. Thus, *proximal* goals whose attainment is projected in the near future can be distinguished from *distal* goals whose attainment can only occur in the far future. In that vein, time horizon is a combined function of both behavior orientation (expressive versus instrumental) and goal orientation (proximal versus distal). Assuming both subdimensions as being dichotomous, four types of time horizon emerge. Figure 1 (p. 106, which is adopted from Jones, 1988, p. 35 in a simplified and slightly modified manner) schematically represents these four types.

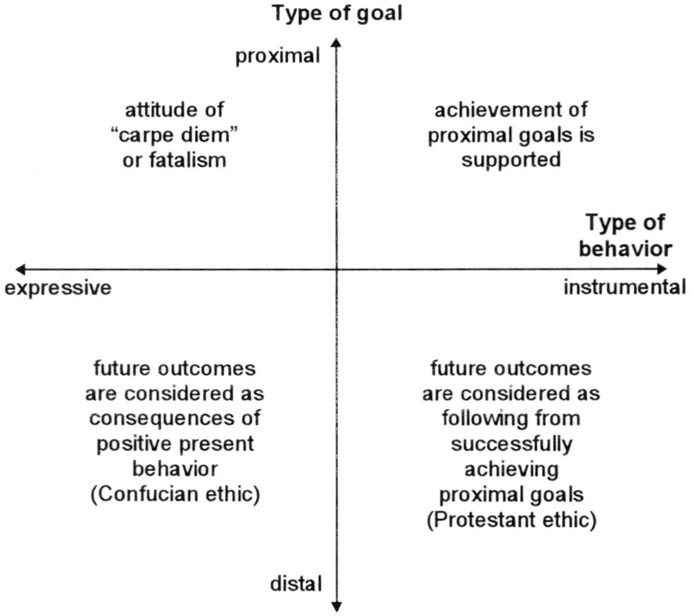

Figure 1: A model of time horizon (modified after Jones, 1988, p. 35)

An extremely short time horizon (upper left quadrant of Figure 1, p. 106) is characterized by having proximal goals and acting expressively, representing either an attitude of „carpe diem“ or of fatalism. An extremely long time horizon (lower right quadrant of Figure 1, p. 106) is characterized by both distal goals and behaviors which are instrumental with respect to these distal goals. It should be noted, that, being extremely future-orientated, may not be confounded with rationality of behavior. A future-driven temporal perspective is rational only when the individual has received confirming evidence about the conditional probability that present action will alter future outcomes. Hence, in some cultural contexts, favoring proximal goals and performing acts which are instrumental with respect to these proximal goals (upper right quadrant of Figure 1, p. 106) may be more rational than having a purely future perspective. For example, it has been found (Bochner & David, 1967; quoted according to Jones, 1988, p. 30) that aboriginal Australian children who favored proximal goals were more intelligent than those who favored distal goals. The interpretation of these results was that the more intelligent children had a more correct estimation of the uncertainty of the future than the less intelligent children.

The type of time horizon associated with the lower right quadrant in Figure 1 (p. 106) is typical for Western, i.e. American-European cultures. Two interrelated reasons have been identified to account for the extreme future orientation of Western societies. The first is the Protestant work ethic with its orientation to individual achievement and success. According to this ethic, the individual in Western societies has to learn that he or she has to delay present gratification in order to attain future goals and that he or she is responsible for the successful or respectively unsuccessful achievement of these future goals. The radical pursuit of the norm of individual responsibility for success in life has led to the consequence that the entire biography has to be mastered individually. This „individualization“ of biography (cf. Beck, 1986) is the second reason for the extreme future-oriented perspective in Western cultures. In contrast to traditional societies, the global structure of the individual biography in Western societies is far less prefigured with the consequence that the individual has to develop coping strategies to reduce future uncertainty and has to realize that both success and failure will be attributed to him or herself and not to external circumstances.

At first glance, the combination of expressive behavior with distal goals, as represented by the lower left quadrant in Figure 1 (p. 106), seems to imply a contradiction. However, the contradiction is resolved insofar as (according to that pattern) maintaining positively valued behavior in the present is viewed as securing a positive future (cf. Jones,

1988, p. 35). This type of time horizon seems to be typical for Eastern societies with a Confucian ethic. Although the goals are located far in the future, the present behavior has its own intrinsic value and does, at least from the Western perspective, not bear an apparent instrumental relationship to the future goals. Thus, maintaining harmonic social relationships is more highly valued than a directly goal-orientated behavior. However, there is a strong belief, that maintaining harmonic relationships in the present secures a harmonic future.

Applied to the interindividual level, the four types of temporal perspective are not mutually exclusive. Thus, for example, even in a society dominated by Protestant work ethic, not all individuals favor the future-driven time horizon. Moreover, even at the intrapersonal level, future and present time horizon are not opposites, but rather are domain-specific: for most Western people, workdays are dominated by the future time horizon, while weekends and evenings permit some degree of a present time horizon.

Social time

While time horizon refers to the individual projection of life time, the social organization of time refers to temporal rules and regulations in social life. Obviously, subjective and social time are mutually dependent, however, one can assume that the interaction is asymmetric, insofar as the degree of independence of social prefigurations is higher in the Western culure than in traditional cultures.

Two aspects of social time organization may be discerned: the social *allocation of time* and the *social pace of life*. There is no consistent correlation between the two aspects across cultures: while the social pace of life seems to depend on global cultural features such as industrialization or urban versus rural environment, time allocation seems to be contingent on the dichotomy of Western versus non-Western cultural views.

Social allocation of time

Cultures differ in their norms for what constitutes punctuality and lateness, and in the meaning of „now“ and „later“. Hall (1959) describes these temporal rules as the „silent language“ of a culture, and distinguishes a „monochronic“ from a „polychronic“ time pattern (cf. Hall & Hall, 1985). The former, characterized by punctuality, time pressure and exact time schedules is typical for Western „individualistic“ cultures with their emphasis on goal-related performance (cf. Hofstede, 1980; 1993). The latter allows one to ignore fixed time schedules and is typical for the need for social harmony in „collectivistic“ cultures (cf. Bond, 1994; Yamaguchi, 1994).

Not all cultures can be consistently assigned to either the polychronic or the monochronic time pattern, sometimes the preferred orientation is domain-specific. Perhaps most apparent is the differentiation according to social domains in Japan: while technical contexts and encounters with foreigners are characterized by punctuality and monochronic time allocation, other domains of life and work are dominated by polychronic time allocation. For instance, a conference meeting starts punctually, but once started, time schedules are of minor importance (cf. Hall & Hall, 1985, p. 99). One might be tempted to suggest an „outer“ time in comparison to an "inner" time in the case of Japan. The „outer“ time reflects Western industrialization, while the „inner“ time has its roots in the traditional value of social harmony typical for collectivistic societies. However, the „timelessness“ associated with polychronic time allocation is by no means typical just for Japanese or collectivistic societies in general; it can also be observed in Western companies, though here it is restricted to the higher ranks of the hierarchy (cf. Deutschmann, 1982).

Social pace of life

Cultural norms also exist for appropriate speeds in work, in play, and in other actions. It has been argued that, adjusting to an alien social tempo may be as difficult as learning the foreign language itself (cf. Levine, this volume, chapter 7). Levine & Wolff (1985; cf. Levine, 1988) studied the social tempo, the so-called *pace of life* in large and medium-sized cities in six countries (England, Italy, Indonesia, Japan, Taiwan and the USA) in terms of three indicators: the average walking speed of pedestrians, the accuracy of bank clocks, and the speed with which postal clerks fulfilled a standard request for stamps. There were strong and consistent differences in all three measures among the individual countries: the highest social tempo was found in Japan, the lowest in Indonesia (see Levine, this volume, chapter 7).

Biological time

Biological time of human organisms may be studied on both the macro and micro level. The macro level involves epochal and certain irreversible biological events such as birth, sexual maturation or death, while the micro level refers to recurrent cycles such as activity-rest, sleep-wake, heartbeat or temperature cycles. As we shall see, temporal features are affected by cultural factors on both levels. To characterize these cultural influences, I use the term „entrainment" (McGrath & Kelly, 1986; Kelly, 1988). The term was originally coined in the biological sciences to refer to the influence which is exerted on one endogenous rhythmic process by another internal or external rhythmic process (Kelly, 1988, p. 89). When a rhythmic process becomes strongly entrained, it is modified in periodicity or phase (or both) toward the entraining cycle. Analogously, the term „social entrainment" has been used to describe social influences on human „natural" activities (Kelly, 1988, p. 90).

Epochal biological events

Epochal events in biological life are birth and death as well as sexual maturity, becoming a mother or father, menopause, getting ill and growing old. Some of these „events" can be fixed at a certain point in time (such as birth and death), others extend as processes over a longer time period (such as growing old). Although the extent of susceptibility to cultural and social influences varies, most of these biological events and processes can be and are culturally modified. For instance, what is judged as „premature" in juveniles, depends less on biologically defined statistical norms but rather on norms of the cultural environment, thus modifying subsequent juvenile behavior and biological states (cf. Silbereisen et al., 1993).

In almost all cultures, there exist „rites of passage" or initiation rites which mark the transition from childhood to youth. These rites are not restricted to traditional cultures, but have survived in highly industrialized societies as religious ceremonies and even in atheistic societies such as the former German Democratic Republic where they have been replaced by the secularized rite of the „Jugendweihe." The chronological point, at which these rites of passage take place, has consequences on subsequent behavior, such as the use of alcohol or engaging in sexual behavior, and thus shapes subsequent biological states.

The effect of cultural influence on biological processes can be easily demonstrated in the case of the chronological age when adults (or juveniles) become parents. For instance, Germany before reunification: a comparison between the former German Democratic Republic and the Federal Republic of Germany of the age of becoming a parent shows that the former had a much lower average than the latter. The reason for that was the social gratification associated with the status of parentship: while in the Federal Republic children were a hindrance with respect to career and life style, legal regulations of the German Democratic Republic offered gratification such as additional housing space and money.

Biological cycles and rhythms

There are a number of biological processes which form periodically repetitive events. Examples of such rhythms are heartbeat, breathing or the sleep-wake cycle. Most evident is the existence of a *circadian* rhythm with a single period including approximately a day. Compared with lower order biological species, the circadian rhythm of humans is less affected by the entraining power of the natural changes in darkness and in temperature that accompany the earth's daily rotation: even under absence of external stimuli such as sun position and brightness, that is, under so-called free-running conditions, the duration of the cycle is approximately conserved. The human species has developed elaborate temperature regulatory systems that function to maintain an internal temperature cycle quite distinct from ambient temperature (cf. McGrath et al., 1984). This temperature cycle, under normal conditions, is synchronized with overall activity levels, and with many other physiological cycles as well. Such synchronization, or lack of it, has dramatic implications for much of human health and well-being, as well as for effectiveness of task and work performance (cf. McGrath et al., 1984).

Surprisingly, the duration of the cycle may be strongly affected by socially mediated time markers such as gong beats having the function of action requests. Within certain limits, such social cues may cause the „internal day" to shrink or lengthen for hours. The socially lengthened day subjectively contains the same number of hours as the socially shortened one: there is a strong, positive correlation between one-hour time estimates and the socially entrained day (cf. Aschoff, 1986). That means that time is subjectively extended or compressed. From those studies we may conclude that humans are strongly entrained to a culturally defined cycle of day and night (and to other culturally defined cycles as well) rather than to an astronomically defined cycle.

Cultures may differ with respect to the extent of external entrainment they impose on the individuals' rhythms. Cultural environment may enhance or attenuate the synchronization of biological rhythms. An example for enhancement would be meditation or certain religious practices, an example for attenuation or even distortion are certain work schedules involving night work or rotating shifts.

Entrainment is not restricted to the modification of physiological processes but also extends to behavior cycles such as performance rate in working situations. Many studies (cf. McGrath et al., 1984; Kelly, 1988) have shown, that person-specific, „natural" rates of completing given tasks can be persistently altered by external pacers. Such an external pacer is a time limit given for the work to be performed. Individuals increase their work rate when time is short and decrease their work rate when time is abundant (Kelly, 1988). Time limits also affect the perceived difficulty of the task. When time limits are either too short or too long for the amount of work, the task is judged to be more difficult than when time limits are appropriate to the natural working rate.

Such an adaptation to external pacers also occurs on a group level. When time is short, work groups cut down all nontask behaviors such as interpersonal interactions, and, when time is abundant, they engage in more personal interaction. However, there is a trade-off between rate and quality: short time limits induce and entrain high rates of performance, but at the same time the quality of performance is lowered, whereas long time limits induce and entrain interaction patterns that lead to lower rates but higher quality of performance (Kelly, 1988).

Cultural differences in short interval estimation

We may ask whether cultural differences in temporality affect the perception of time. As the studies on circadian rhythms and their entrainment by social pacers show, estimation of time intervals in hours elapsed are strongly affected by circadian rhythms whether natural or socially entrained. However, entirely different mechanisms seem to be used for the perception of time lapses in the order of minutes and seconds (cf. Groos & Daan, 1985).

Everyday experience shows, that in the absence of external time markers, such as clocks, the estimation of short time intervals strongly depends on the focus of attention: time flies when a task absorbs our full attention, but seems to stand still when the task is boring (cf. Funke & Grube-Unglaub, 1991). In retrospect, the reverse takes place: the accelerated time is subjectively extended and the time slowed down is subjectively shrunk. To explain this time paradox, it is suggested that a time period filled with a demanding task imprints more and richer traces in memory, whereas periods where nothing or little happened, will leave only a few traces (cf. Friedman, 1990, and Block & Zakay, this volume, chapter 9).

As the evaluation of quality and quantity of task demands is strongly shaped by cultural factors, it seems to be justified to systematically study cultural factors such as the social tempo or the conception of time with respect to their influence on time estimations. From a comparison of urban with rural people in Jamaica, Doob (1960) concluded that the heightened social tempo of urban life leads to an enhancement of temporal awareness, but at the same time to less precise time estimations.

Our own research interest has focused on the study of the estimation of time intervals in the range of milliseconds. In everyday life, distinctions within that range are required in order to produce and understand running speech (see also Semjen, this volume, chapter 2). It might be hypothesized that the judgment of two consecutive short events as either both occurring „simultaneous“ or the second „successive“ to the first, may be dependent on the prevalent social tempo or social pace of life. In our study, we compared two samples of students, one recruited from Tokyo, and the other recruited from Regensburg, a middle-sized town in Germany. We presupposed that the social tempo is higher in Tokyo than in Regensburg (see Levine, this volume, chapter 7). Since a higher social tempo implies, that the period of „nowness“ (i.e. a single moment in present time) is shortened, and the period of „lateness“ (i.e. the next moment in time) will start earlier, compared to a lower social tempo, where „nowness“ has a longer duration, and „lateness“ will start later. We expected that the Japanese students in Tokyo would detect the asynchrony between two brief tonal events earlier, i.e. at a shorter time gap, when compared with the German students in Regensburg.

In standard experimental settings on time estimation in the milliseconds range the subjects are asked to indicate whether two brief acoustic events such as sounds or noises have started either simultaneously or successively. According to Pisoni (1977) and Hillenbrand (1984) a time lag of about 20 ms is required, in order to perceive the two onsets as successive.

In our study, conducted with the Japanese students at Sophia University, Tokyo, and with the German students at the University of Regensburg (cf. Helfrich et al., 1991), we had the subjects listen to sound patterns consisting of two tonal components. The tonal components either started simultaneously, or the second component was successive to the first. The time lag between the two onsets, i.e. the onset asynchrony, was systematically varied from 0 ms (synchronous onsets) to 50 ms, in 10-ms steps. Two different types of tasks were applied. In the labeling task, subjects had to decide whether onsets of the two tonal components of a stimulus presented at one time had occurred either „simultaneously“ or „successively.“ In the identification task, subjects had to identify each stimulus (consisting of two tonal components) with a number ranging from „0“ (simultaneous onsets) to „6“ (onset asynchrony of 50 ms). After a practice session, the stimuli (each totally 240 times) were presented in randomized order within blocked tasks (labeling versus identification task).

The labeling data reflect the subjective evaluation of stimuli with respect to asynchrony. They are not bias-free, because what is labeled as

„successive", depends on the subjective criterion of the judging subject. A method to eliminate subjective response bias in order to get a true or bias-free measure of perceptual discriminability of asynchronous from synchronous stimuli is provided by signal detection theory (cf. Macmillan & Creelman, 1991). Thus, the identification data were analyzed using signal detection theory. As a measure of perceptual discriminability between the physically synchronous and the physically asynchronous stimulus, the perceptual discrimination distance d' has been computed. It represents the distance between the sensory means of two stimuli measured in units of their standard deviations (cf. McNicol, 1972, p. 57).

The following results were obtained. When the stimuli are to be labeled as either „simultaneous" or „successive", the critical time interval where the shift from the dominance of „simultaneous" responses to the dominance of „successive" responses (i.e. the 50% cut-off point) takes place is significantly shorter for the Japanese compared to the German subjects ($t = 3.85$; $df = 10$; $p_{two\text{-}tailed} < 0.001$): this shift occurs at an onset asynchrony of 5 ms for the Japanese subjects and at an onset asynchrony of 17 ms for the German subjects (Figure 2, p. 116).

However, the discrimination distance d' between the physically synchronous and the physically asynchronous stimulus based on signal detection theory (Figure 3, p. 116) shows no significant difference between the two groups (for the German subjects the mean discrimination performance even slightly exceeds that for the Japanese).

Although it seems to be premature to generalize from the findings, some preliminary conclusions might be drawn. Thus, we may state, that the social pace of life is contingent with the cognitive evaluation of short-time events, but does not affect the basic perceptual process of temporal differentiation, which might reflect the sensation of „pure time" (cf. Lejeune & Richelle, this volume, chapter 8) and seems to be resistant to cultural influences.

Furthermore, one might be tempted to speculate that the labeling judgments reflect the amount of attention allocated to time (see Block & Zakay, this volume, chapter 9), with a higher amount of attention leading to an earlier shift from „simultaneous" to „successive" judgments. The amount of attention allocated to time seems to be highly susceptible to cultural influences such as the social pace of life.

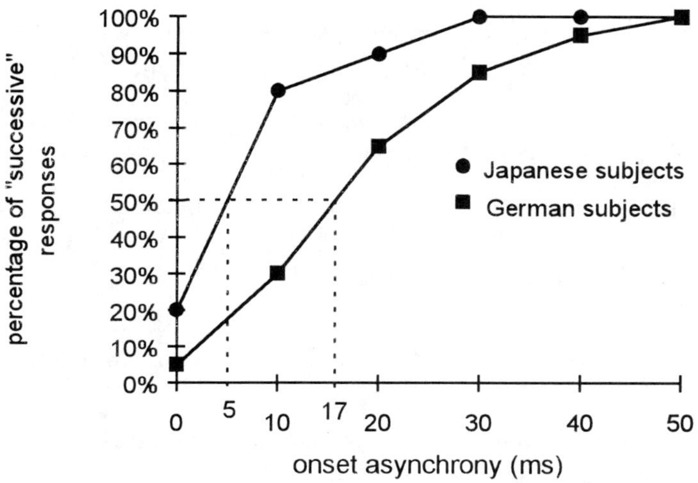

Figure 2: Percentage of „successive" responses as a function of onset asynchrony between the two tonal components of a stimulus for Japanese and German subjects[2]

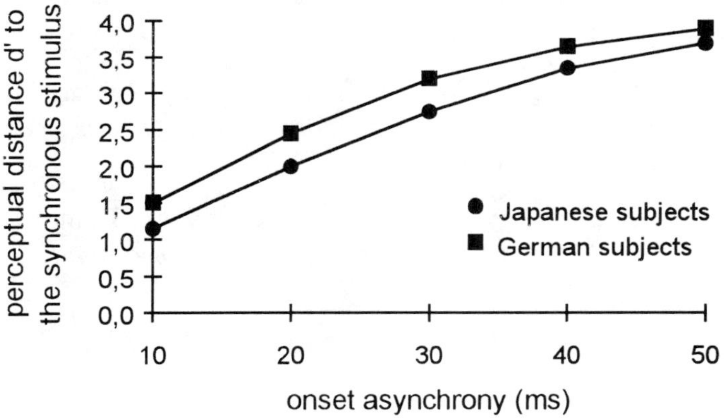

Figure 3: Perceptual discrimination distance d' to the synchronous stimulus as a function of onset asynchrony for Japanese and German subjects[3]

---

[2] The dotted lines mark the 50% cut-off point, where the shift from the dominance of „simultaneous" to the dominance of „successive" responses takes place.

[3] d' is defined as $(\mu_i - \mu_0) / \sigma$, where μ_0 refers to the synchronous stimulus, μ_i to the comparison stimulus, and σ to the respective standard deviations.

References

Aschoff, J. (1986). Circadiane Systeme. In B. Bromm & D.W. Lübbers (Eds.), *Physiologie aktuell. Bd. 2* (pp. 5-15). Stuttgart: Thieme.

Beck, U. (1986). *Risikogesellschaft. Auf dem Weg in eine andere Moderne*. Frankfurt (Main): Suhrkamp.

Bond, M.H. (1994). Into the heart of collectivism: A personal and scientific journey. In U. Kim, H.C. Triandis, C. Kagitcibasi, S.-C. Choi, & G. Yoon (Eds.), *Individualism and collectivism* (pp. 66-76). Thousand Oaks, CA.: Sage.

Deutschmann, K. (1982). *Arbeitszeit und Freizeit in Japan*. Tübingen: Niemeyer.

Doob, L.W. (1960). *Becoming more civilized: a psychological explanation*. New Haven: Yale University Press.

Durkheim, E. (1898). Représentations individuelles et représentations collectives. *Revue de Metaphysique et de Morale*, *6*, 273-302.

Eliade, M. (1953). *Der Mythos der ewigen Wiederkehr*. Düsseldorf: Diederichs.

Friedman, W.F. (1990). *About time: inventing the fourth dimension*. Cambridge, Mass.: MIT Press.

Funke, J. & Grube-Unglaub, S. (1991). Verarbeitungstiefe als Einflußfaktor auf das subjektive Zeitempfinden. *Schweizerische Zeitschrift für Psychologie*, *50*, 129-138.

Groos, G. & Daan, S. (1985). The use of biological clocks in time perception. In J.A. Michon & J. Jackson (Eds.), *Time, mind, and behavior* (pp. 65-74). Berlin: Springer.

Hall, E.T. (1959). *The silent language*. New York: Doubleday.

Hall, E.T. & Hall, M.R. (1985). *Verborgene Signale*. Hamburg: Gruner & Jahr.

Helfrich, H., Faßmann, W., Burmester, M., & Linz, M. (1991). *Kategoriale Lautwahrnehmung*. Regensburg: DFG-Report

Hillenbrand, J. (1984). Perception if sine-wave analogs of voice onset stimuli. *Journal of the Acoustical Society of America*, *75*, 231-240.

Hofstede, G.(1980). *Culture's consequences*. Beverly Hills, CA: Sage.

Hofstede, G.(1993). *Interkulturelle Zusammenarbeit*. Wiesbaden: Gabler.

Jones, J.M. (1988). Cultural differences in temporal perspectives. In J.E. McGrath (Ed.), *The social psychology of time* (pp. 21-38). Newbury Park, CA: Sage.

Kelly, J.R. (1988). Entrainment in individual and group behavior. In J.E. McGrath (Ed.), *The social psychology of time* (pp. 89-110). Newbury Park, CA: Sage.

Levine, R.V. (1988). The pace of life across cultures. In J.E. McGrath (Ed.), *The social psychology of time* (pp. 39-60). Newbury Park, CA: Sage.

Levine, R.V. & Wolff, E. (1985). Social time: The heartbeat of culture. *Psychology Today*, March, 28-35.

McGrath, J.E. & Kelly, J.R. (1986). *Time and human interaction*. New York: Guilford Press.

Macmillan, N.A. & Creelman, C.D. (1991). *Detection theory: A user's guide*. Cambridge: Cambridge University Press.

McGrath, J.E., Kelly, J.R., Machatka, D.E. (1984). The social psychology of time: entrainment of behavior in social and organizational settings. *Applied Social Psychology Annual*, *5*, 21-44.

McNicol, D. (1972). *A primer of signal detection theory*. London: George Allen & Unwin.

Moscovici, S. (1981). On social representations. In J.P. Forgas (Ed.), *Social cognition* (pp. 181-209). London: Academic Press.

Nakamura, H. (1968). Time in Indian and Japanese thought. In J.T. Fraser (Ed.), *The voices of time* (pp. 77-91). New York: Braziller.

Pisoni, D.B. (1977). Identification and discrimination of the relative onset time of two-component tones: Implications for voicing perception in stops. *Journal of the Acoustical Society of America*, *61*, 1352-1361.

Richelle, M., Lejeune, H., Perikel, J.-J., & Fery, P. (1985). From biotemporality to nootemporality: Toward an integrative and comparative view of time in behavior. In J.A. Michon & J. Jackson (Eds.), *Time, mind, and behavior* (pp. 75-99). Berlin: Springer.

Silbereisen, R.K., Schwarz, B., & Kracke, B. (1993). Problemverhalten Jugendlicher im Kulturvergleich. In A. Thomas (Ed.), *Kulturvergleichende Psychologie* (pp. 339-357). Göttingen: Hogrefe.

Thomas, A. & Helfrich, H. (1993). Wahrnehmungspsychologische Aspekte im Kulturvergleich. In A. Thomas (Ed.), *Kulturvergleichende Psychologie* (pp. 145-180). Göttingen: Hogrefe.

Yamaguchi, S. (1994). Collectivism among the Japanese: A perspective from the self. In U. Kim, H.C. Triandis, C. Kagitcibasi, S.-C. Choi & G. Yoon (Eds.), *Individualism and collectivism* (pp. 175-188). Thousand Oaks, CA: Sage.

Chapter 7: Cultural differences in the pace of life

ROBERT V. LEVINE

„It is by the meaning that it intuitively attaches to time that one culture is differentiated from another."

Oswald Spenger (*The Decline of the West, Volume 1*)

Introduction

Cultures may differ on many aspects of social time - its value, its meaning, how it should be divided, allocated and measured. This chapter focuses on perhaps the single most significant distinguishing factor in the social time of cultures, the pace of people's lives. Differences in pace permeate nearly all of day-to-day social activity. The pace of life is virtually inseparable from the entire experience of social time, which in turn frames the totality of both our deepest relationships and our personal experience. The shape of social time underlies the entire configuration of our social relationships and these patterns define much of the uniqueness of cultures.

It has been argued that adjusting to the pace of life in a foreign country can be as difficult as learning the foreign language. In an investigation into the roots of culture shock, Spradley & Phillips (1972) asked 83 returning U.S. Peace Corps volunteers to rank in order 33 items as to how much difficulty they posed for cultural readjustment. Their list included issues ranging from „the type of food eaten" to the „personal cleanliness of most people, „the number of people of your own race," and „the general standard of living." But, after „the language spoken," the two stressors that posed the greatest adjustment difficulties to the volunteers both related to temporal matters: „the general pace of life" followed by „how punctual most people are." Howard (1980) reported that overseas business managers and their families experienced similar problems when returning home from slower paced cultures.

Despite the profundity of cultural differences in the pace of life, few systematic studies have measured these differences. Previous research on cultural differences in the pace of life has mostly taken an anthropological, case study approach compiling broad, often subjective, descriptions of specific behaviors and norms in a given culture. These studies have produced a wealth of case by case profiles of the pace of life and overall social time structures from a diversity of cultures.

The shortcoming of these studies, however, is their lack of quantitative data about a sufficient number of subjects to allow systematic cross-cultural comparisons. Without suitable comparison groups, it is difficult to gauge the meaning of „fast" or „slow" or to grasp the implications of one people's time constructions from another's.

McGrath (1989) has observed how remarkably little attention the study of time has received from social psychology. Nowhere is this more apparent than in pace-of-life studies. With the exception of a few studies on urban-rural differences, there is virtually no reported research which has taken a broadly based social psychological-type approach to the topic (see the works of Bornstein and his colleagues, e.g. Bornstein, 1979, and Amato, 1983, for some notable exceptions; see Levine, 1988, for a review of cross-cultural pace of life studies). With this consideration in mind, my colleagues and I have carried out a series of cross-cultural and within-cultural studies over the last fifteen years.

Study 1: The pace of life in six sountries

Our first studies of the pace of life focused on the largest city and one medium-sized city (population between 415,000-615,000) in each of six countries: Taipei and Tainan (Taiwan), Tokyo and Sendai (Japan), Jakarta and Solo (Indonesia), Rome and Florence (Italy), London and Bristol (England), and New York City and Rochester, N.Y. (United States). This sample allowed for comparisons between Eastern and Western cultures varying in degree of economic development. To control for seasonal temperature variations, measurements were made during either the summer or another warm weather month. All measures were taken in downtown areas during main business hours and at approximately the same time of day in each city.

We examined three indicators of tempo in each city. First, as an estimate of a city's interest in clock time, we observed the accuracy of 15 randomly selected bank clocks in main downtown areas in each city. Times on the 15 clocks were compared to those reported by the telephone company. Deviations from the correct time were measured to the nearest minute.

Second, we measured the average walking speed of randomly chosen pedestrians over a distance of 100 feet. Measurements were taken during core business hours on clear days in at least two locations on main downtown streets. Locations were chosen that allowed pedestrians to move at potentially maximum speeds. In order to control for the effects of socializing, only pedestrians walking alone were used. Neither subjects with clear physical handicaps, nor those who appeared to be window-shopping were timed.

Third, as a measure of work pace we measured the speed with which postal clerks fulfilled a standard request for postage stamps. In each city, including those in the United States, we presented clerks with a note in the native language requesting a common-priced stamp - the then standard 20-cent stamp in the United States, for example. They were also handed paper money - the equivalent of a $5 bill. We measured the elapsed time between the passing of the note and the completion of the request.

Our results indicated a number of consistent cross-national and within-national differences. There were significant main effects for the factor country on all three measures. As seen in Figure 1 (p. 122), the Japanese were highest on all measures. They had the most accurate clocks, the fastest walking speeds and the fastest postal clerks. The United States was second on the bank clock and post office measure and slightly, but significantly, slower than Japan on walking speed (22.5 seconds versus 20.7 seconds).

On the tardy end, Indonesia had the least accurate clocks and the slowest walking speeds by far. When it came to efficiency of postal service, however, the Indonesians were not the least efficient (although they did place far behind the Japanese postal clerks). That distinction went to the Italians, whose infamous postal service took 47 seconds on the average (compared to Japan's average of 25 seconds).

There were also a number of within-country differences, including a country x city size interaction for walking speed. As expected, we found here that people in the larger cities walked faster than those in the smaller cities. But the interaction indicated that this difference was greatest in the least developed countries in our sample - Indonesia followed by Taiwan. Perhaps, we hypothesized, this reflects the persistence of a traditional village life style among people in the smaller cities in Indonesia and, to a lesser extent, Taiwan. The traces of a traditionally rural slow pace-of-life may be less evident, or never existed in some cases, in the smaller cities of the more industrialized nations.

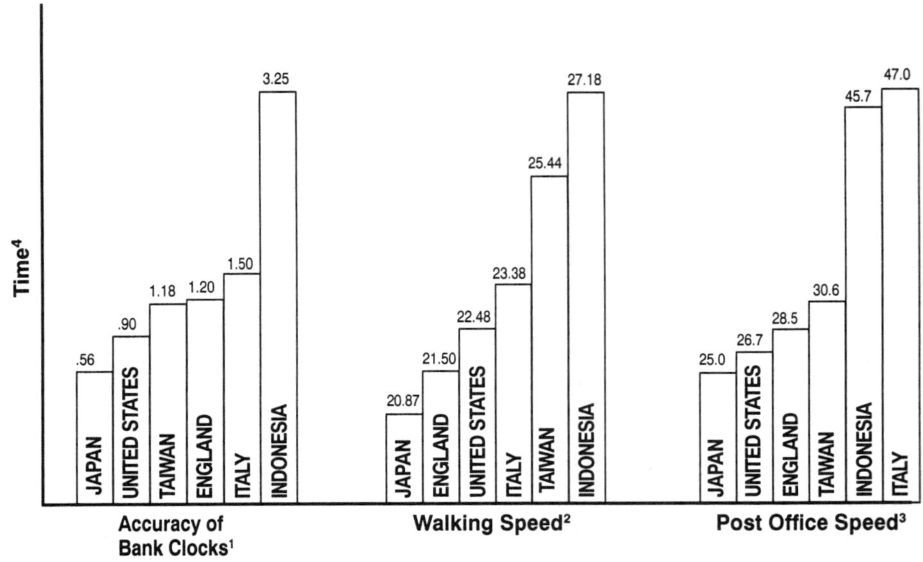

1 Deviations are reported in minutes

2,3 Speeds are in seconds

4 Smaller numbers indicate more accurate clocks, faster walking speeds and faster office speeds respectively

Figure 1: The pace of life in six contries (Levine, 1988)

One of the strongest findings to come out of these studies was the consistency of the three pace of life measures across the twelve cities in our sample. Accuracy of clocks was positively related to faster walking speeds (r = .82) and to faster completion of postal requests (r = .71). Walking speed was positively correlated with postal efficiency (r = .56). These data provided strong evidence for the stability and generality of the pace of life concept.

In summary, these results demonstrated that there are strong and consistent differences among nations in their paces of life, and that these differences are relatively stable across measures. It appears reasonable to speak of a characteristic pace of life of a particular people and to distinguish between cultures on the basis of this characteristic. (See Levine & Bartlett, 1984, and Levine, 1990, for a complete description of these studies).

Study 2: The pace of life and coronary heart disease

Our next studies focused on the consequences of the pace of life. Is there a relationship between the tempo of an area and the psychological and physical health of its people? (See Levine et al., 1989, for a complete description of these studies).

Our research team focused on coronary heart disease (CHD). Extremes in personal tempo have long been suspected risk factors for coronary heart disease. A chronic sense of time urgency is a defining component of the type A syndrome - which is also characterized by hostility, aggression and competitive achievement striving (e.g. Friedman & Rosenman, 1974). Since time urgency represents one extreme on the pace of life continuum, we predicted that the faster a city's overall pace of life, the higher its rate of coronary heart disease would be.

Our studies come at a time when the validity of the entire type A pattern is being disputed. Several recent studies have argued that the only type A element that actually relates to coronary heart disease is „hostility-anger." These studies have minimized the significance of time urgency (see reviews by Booth-Kewley & Friedman (1987) and Matthews (1988). Other researchers, however, have found that time urgency, along with the related tendency to remain chronically active and 'keyed up', may indeed be a contributing factor to heart disease (Wright, 1988).

Is time urgency a lethal type A ingredient? Previous investigations have relied on self-reports, through questionnaires and interviews, to answer this question. Rather than studying time urgent personalities, we focused on time urgent environments. Is there a relationship, we asked, between an area's characteristic pace of life and its rate of CHD?

Our six-country study revealed that reliable international health statistics are hard to come by (Reid, 1975). This is especially true at the level of cities. To make generalizations about the relationship between pace and other community characteristics would require a considerably larger sample of cities than we tested. Rather than studying foreign cultures for this larger sample, we decided to first focus on the United States.

Our „subjects" were 36 cities from across the United States. To achieve a representational sample of U.S. cities, we studied three large cities (metropolitan area populations above 1,800,000), three medium-sized cities (populations between 850,000-1,300,000) and three smaller ones (populations between 350,000-550,000) from each of the four census-defined regions of the United States (the northeast, north central, south and west).

We examined four indicators of the pace of life in each city. First, we observed walking speed. We clocked how long it took pedestrians to walk 60 feet along main downtown streets during business hours on a clear summer day. We applied the same restrictions on selection of subjects and locations as in the six-country study.

Second, as an indicator of work speed, we timed how long bank clerks took to complete a simple request. We alternately asked tellers in each of at least eight downtown banks either to give change, in set denominations, for two $20.00 bills or to give us two $20.00 bills for change.

Third, talking speed was observed. In each city we tape recorded the responses of postal clerks to a standard question - to explain the difference between regular mail, certified mail and insured mail. We then calculated their 'articulation' rates by dividing the number of syllables in their response by the total time of the response.

Fourth, as a simple measure of concern with clock time, we counted the proportion of men and women in downtown areas during business hours who were wearing wristwatches.

Individually, each of these measures has its quirks - they all tap into special sub - populations, the second two are confounded by skill and efficiency, and the last can be affected by fashion styles. However, taken together, they sample a wide range of people and activities, and reflect many facets of a city's pace of life.

Finally, we created an index of the overall pace of life of each city by statistically standardizing each of the four scores - so they carried equal weight in the final index - and adding them together. The results can be seen in Table 1 (p. 125).

The three fastest cities, and seven of the first nine, were in the Northeast. Overall, Easterners walked faster, gave change faster, talked faster and wore more watches than their compatriots in other U.S. cities.

Boston, Buffalo and New York City were the fastest. Although New Yorkers were slightly slower than subjects in the other two cities, they might be excused for a couple of slow steps to watch the local festivities. Walter Murphy, who collected the walking speed data there, reported an improvised music concert, an attempted purse snatching, and a mugger during the hour and a half that he was clocking pedestrians on one corner alone.

Pace of Life measure												
R = rank M = mean	overall index		walking speed		bank speed		talking speed		watch frequency		CHD/ median age	
City	R	M[1]	R	M[2]	R	M[3]	R	M[4]	R	M[5]	R	M[6]
Boston, MA	1	53.36	2	11.34	2	40.97	6	3.46	2	85.0	10	9.3
Buffalo, NY	2	49.63	5	11.54	7	41.40	4	3.50	4	83.0	2	11.2
New York NY	3	32.77	11	12.31	11	44.75	18	3.01	1	90.0	1	12.1
Springfield, M	4	30.24	1	11.09	15	46.89	7	3.45	22	68.5	7	10.0
Providence, RI	5	25.08	7	11.94	9	43.77	8	3.41	18	70.0	3	10.3
Rochester, NY	6	22.98	20	12.67	2	35.30	16	3.03	7	76.5	14	7.9
Columbus, OH	7	21.55	22	12.87	17	48.08	2	3.80	18	70.0	26	6.7
Bakersfield, CA	8	20.88	28	13.67	13	45.84	1	3.93	17	71.5	20	7.4
Patterson, NJ	9	18.50	17	12.66	4	37.89	3	3.60	31	61.5	4	10.2
Canton, OH	10	14.18	23	13.08	14	46.83	5	3.48	14	72.0	9	9.4
Worcester, MA	11	12.40	9	12.02	22	50.89	22	2.94	6	77.0	5	10.2
St. Louis, MO	12	11.94	15	12.51	20	50.10	10	3.30	14	72.0	8	9.6
Salt Lake City, UT	13	11.17	4	11.51	16	47.60	29	2.73	10	73.5	31	5.5
Houston, TX	14	8.88	10	12.20	8	43.53	23	2.92	18	70.0	36	4.2
Kansas City, MO	15	7.47	6	11.81	3	36.19	25	2.88	32	60.0	21	7.3
Detroit, MI	16	3.82	21	12.81	12	45.28	36	2.39	2	85.0	11	9.2
Philadelphia, PA	17	-.36	30	13.84	5	40.30	20	2.95	10	73.5	16	7.7
Atlanta, GA	18	-2.64	3	11.50	19	55.41	10	3.30	36	55.0	33	5.0
Knoxville, TN	19	-2.79	25	13.17	24	52.26	15	3.04	10	73.5	17	7.6
Indianapolis, IN	20	-4.25	17	12.66	23	51.88	18	3.01	24	68.0	22	7.0
San Francisco, CA	21	-4.33	19	12.67	35	69.71	13	3.07	5	81.5	27	6.4
Louisville, KY	22	-6.80	16	12.58	21	50.33	31	2.66	14	72.0	18	7.5
Chattanooga, TN	23	-7.43	35	14.61	1	32.70	20	2.95	24	68.0	12	8.9
Oxnard, CA	24	-10.10	30	13.84	30	56.03	16	3.03	7	76.5	34	4.5
Youngstown, PA	25	-17.16	13	12.49	18	49.25	33	2.64	30	62.0	6	10.2
San Diego, CA	26	-17.32	27	13.43	34	66.81	13	3.07	9	75.0	24	6.8
Memphis, TN	27	-17.60	34	14.20	10	48.93	9	3.32	24	56.5	30	5.7
Nashville, TN	28	-20.88	8	11.97	26	53.95	32	2.65	33	58.0	15	7.9
Fresno, CA	29	-21.56	36	14.73	25	53.01	12	3.13	18	70.0	25	6.8
Chicago, IL	30	-26.56	12	12.48	31	57.61	35	2.47	27	65.0	13	8.1
San Jose, CA	31	-27.30	29	13.75	29	56.02	30	2.72	22	68.5	35	4.5
Dallas, TX	32	-27.50	26	13.22	28	55.81	28	2.74	28	63.0	32	5.4
Shreveport, LA	33	-27.77	32	14.04	19	49.42	26	2.82	28	63.0	19	7.5
East Lansing, MI	34	-28.91	14	12.50	33	61.88	24	2.91	34	56.5	29	5.8
Sacramento, CA	35	-32.21	33	14.13	32	57.69	26	2.82	26	67.0	23	7.0
Los Angeles, CA	36	-44.54	24	13.09	36	82.41	34	2.61	13	73.0	28	6.3

Table 1: Means and ranks[7] of cities on pace of life and coronary heart disease (CHD)

---

1 T-score with M = 0

2 in seconds

3 T-score with X = 50

4 syllables/second

5 proportion wearing watches

6 death rates * 100,000,
population * median age of population
Based on figures for overall metropolitan area (USA).

7 Lower ranks indicate faster pace and more CHD

The slowest overall pace of life was in the West, due mostly to particularly slow walking speeds and bank tellers. The very slowest times were in Los Angeles. They were 24th in walking speed, next to last in speech rate and far behind everyone in bank speeds. Their only concession to the clock was to wear one - the city was 13th highest in the number of watches worn.

Our main question concerned the relationship between the overall pace of cities and their rates of death from coronary heart disease. Since age is highly related to CHD, and we were only interested in social psychological influences, we first statistically adjusted the CHD figures for the median age of each city's population. Although this seemed to be the most valid age correction, we also later re-analyzed the data after adjusting for the proportion of residents 65 years of age and older.

There was a significant correlation between pace and CHD across the 36 cities - varying from $r = .50$ to $r = .51$, depending on the age correction used. The magnitude of this correlation, in fact, was higher than that usually found between CHD and type A *personality* measures. In other words, our data indicate that the speed of a person's environment is as good a predictor of coronary heart disease as their type A personality test scores. This was true no matter which method we used to correct for age, or whether we took age into account at all.

Why are faster environments more prone to coronary heart disease? In large part, we suspect, because they embrace many type A individuals. Fast places attract fast people. And fast people create fast places. Timothy Smith and his colleagues have shown that type A's both *seek* and *create* time urgent environments. The fastest cities in our study may represent both their dreams and their creations (e.g. Smith & Anderson, 1986).

First, type A's should be attracted to fast paced cities, resulting in more type A residents, who in turn maintain and further encourage fast pace norms. Second, in a similar manner, type B's will tend to migrate away from fast paced cities, to environments more compatible with their temperaments. Third, Smiths' research indicates that the temporal expectations of fast paced cities demands time urgent behavior in all people - A's and B's alike. In the end, type B's act more like A's, and A's strive to push the beat even faster - all in an environment where migration patterns have already created an over-representation of coronary-prone personalities.

The precise mechanism linking time urgent behavior to coronary heart disease is open to speculation. But some recent statistics from the U.S. Department of Health and Human Services (1987) hint at one explanation. Smoking rates follow the same regional patterns as do our data for coronary heart disease and the pace of life. Smoking and CHD rates are highest, and the pace is fastest, in the Northeast, followed by the Midwest, the South and then the West.

Cigarette smoking has been identified as the single most important preventable cause of CHD. It is also well documented that cigarette smoking is often related to psychological stress. We speculate that the stress of time pressures may lead to unhealthy behaviors, such as cigarette smoking and perhaps unhealthy diets, which in turn increase the risk of CHD.

Speed and hostility/anger

Can we conclude from our data that time urgency is a lethal type A ingredient? Or, is it related to CHD only because of an association with hostility/anger? Comparing the present USA findings to those from our earlier cross-cultural study offers some preliminary answers to this question.

In the six-country study, the Japanese were first on every measure: their clocks were the most accurate; they walked the fastest; and their postal clerks on the average took the least time to sell a stamp. To our surprise, however, they had by far the *lowest* CHD death rates of the four countries for which statistics were available (none were available for Taiwan or Indonesia - which, perhaps not coincidentally, were the slowest cities in our sample). Japan's CHD mortality rates, in fact, were the very lowest of 54 industrialized countries compared by the World Health Organization.

How does a type A behaving population avoid CHD? Cultural values - and understanding how the pace of life is embedded in this value system - may provide an answer.

A study by Marmot & Syme (1976) found that Japanese-American men who had experienced non-traditional Japanese cultural upbringings were 2 to 2.7 times more likely than those with traditional upbringings to experience coronary heart disease. These results held even when the traditional CHD risk factors - diet, smoking, cholesterol, blood pressure, triglycerides, obesity, glucose and age - were taken into account.

Perhaps the most highly respected of all Japanese values is social harmony. Competition and aggression have little place in their group-oriented culture. And this may be their buffer against CHD.

The type A personality scale includes a series of questions about being hard-driving-competitive, short-tempered, impatient - as well as a series about being hard working. For U.S. subjects, high scores on the former questions tend to be accompanied by high scores on the latter - all are part of the same behavior pattern. But for Japanese respondents, studies have shown that hard-driving has little to do with being hard working (Cohen et al., 1975).

Even translating 'hard-driving' type A items is a problem. One frustrated researcher reported that the best Japanese translation he could find for the question „Do you like competition on your job?“ was „Do you like impoliteness on your job?.“

It appears that competitive hostility and anger play only a small part in the hardworking, fast pace of the Japanese. In our own culture, though, there is often a fine line between speed and time urgency on the one hand and competition and hostility on the other.

These findings suggest that speed and time urgency may not be lethal in of themselves. They may be related to CHD only when hostility/anger is also present - as was likely the case in our 36 city U.S. sample.

Possibly more important, these data offer a valuable perspective on the way of living in the United States and many other developed Western countries. Workers in the West have often come to view the choice between rushing and leisurely activity as a tradeoff between accomplishment on the one hand and peace of mind on the other. But the Japanese example shows that hard work and productivity need not be destructive to one's health.

Study 3: International differences revisited

Our most recent experiments, which are still in progress, have focused on the pace of life in what is currently a sample of 28 countries from around the world. The 36-city U.S. experiments taught us the value of studying a large number of cities. The sampling of multiple cities in that study had allowed us to treat each city as if it were an individual subject when applying statistical analyses. These correlational analyses revealed a significant consequence of a communities' pace of life, offering city-level insights that might not be easily visible at the individual level. The current experiments have applied this correlational design to the study of other community characteristics that may be related to its general pace of life.

These recent studies have cast a much wider net than the predecessors. Rather than focus on a single consequence of pace, we are now attempting to understand what community traits underlie and result from the pace of life on a broader level. One purpose of the studies has been, as before, to chart the current tempo of a large group of nations. More than that, however, we are concerned with understanding the position and role of a culture's pace of life in the broader web of characteristics that define its personality. Specifically, we asked three main questions.

First, can we identify those community characteristics which best predict differences in the pace of life? Why are some cultures faster than others? We hypothesize that a culture's pace of life is, in fact, the product of a wide range of factors: population demographics, environmental conditions, cultural values and economic conditions. Previous research and popular folk wisdom have suggested that the pace of life is intertwined with three factors: (1) Population size; several previous studies have found that bigger cities tend to be faster (e.g. Bornstein & Bornstein, 1976; Bornstein, 1979); (2) Climate; this is the popular notion that hotter places are slower (although rarely tested, an experiment by Hoel, 1968, did find that walking speed increased as temperature decreased); (3) Individualism-collectivism; although the relationship between individualism-collectivism and social behavior is somewhat complex, previous research would indicate that societies emphasizing individualism should generally have faster paces of life than those which place greater emphasis on the well-being of the collective. Triandis (1994) argues that a major antecedent of individualism is affluence, which is itself both a precursor and consequence of a rapid pace of life (see below). In individualistic cultures, according to Triandis, much behavior occurs when individuals are alone, or interacting with a single other individual. The focus of achievement is also more on individual performance, as opposed to that of the group. Triandis likens much of individualistic culture to what Fiske (1992) describes as a „market pricing“ behavior pattern, under which it is assumed that the more people give, the more they get back in return. These behaviors should be conducive to a rapid pace of life; and (4) Economic vitality; Hoch (1976) and Bornstein (1979), for example, have argued that the more vital a city's economy, the greater the value of people's time, resulting in faster norms.

Second, what is the relationship of a city's pace to its overall well-being? In the U.S. study we found a strong relationship between the pace of life and rates of death from coronary heart disease. As the case of Japan made clear, however, this relationship can be modified by other cultural dynamics. Overall, we now asked, would the relationship between pace and (5) CHD hold up on the larger, and more culturally heterogeneous, international level? Once again, we also examined the relationship of pace to (6) smoking rates, which we found in our earlier study to generally be higher among individuals living in time urgent environments. Finally, and perhaps most important, we investigated an additional indicator of psychological well-being, one which is both broader and more direct: (7) Ratings of life satisfaction. Are people in slower places happier with their lives than their counterparts in type A environments? The lay wisdom of travel brochures would suggest that slower lifestyles are the raw material of happiness. However, there is ample evidence that a nation's economic well-being is highly correlated with its citizen's ratings of life satisfaction (Diener et al., 1994), and one of our central arguments is that faster paces will be reflected in economic achievement. Following the latter reasoning, we predicted that people in faster places would express higher life satisfaction.

The third question focuses on the relationship of the pace of life to social responsibility. Anthropologists have observed that some cultures believe that a rapid pace of life interferes with the carrying out of one's social obligations. The Kelantese, for example, consider haste a breach of ethics. At the core of their sacred *budi bahasa* (the „language of character") is a willingness to allocate the time necessary for social obligations. Rushing signals greed. It is believed to threaten harmonious interpersonal relations and village solidarity (Raybeck, 1992). If the reasoning behind the Kelantese code is correct, we would predict that people in cultures with a faster pace of life are less likely to come to the assistance of others in need. It is noteworthy that contemporary cognitive processing theories use a very different argument to make the same prediction. Milgram's (1970) influential system overload theory, in particular, argues that a rapid pace of life and inattention to the needs of strangers go hand in hand. According to Milgram, when people are confronted with more sensory inputs than they are able to process they experience psychological overload. This is frequently the case in modern cities - the larger the city, the greater the overload. One way that the overloaded urbanite adapts to this predicament is by screening out nonessential stimuli. In essence, the city dweller focuses on his or her goals and moves directly toward them as quickly as possible. Since the needs of strangers are usually low on the urbanite's hierarchy of necessities, attention to these needs become a frequent casualty of the screening process. The third goal of our study, then, was to test the predictions of the *budi bahasa* and Milgram: specifically, (8) are people in fast places less likely to take the time to help a stranger?

We have tested these hypotheses through experiments in a sample of 28 countries. The selection of countries has been aimed at the widest possible sampling of the regions and cultures of the world. Although this goal has been largely accomplished, it is important to note that the representativeness of the sample has been limited by practical considerations. These studies, like those before, have been conducted with minimal financial assistance. Given these limitations, we turned to our own local students for research assistance. Data collection at the various international sites has been conducted, for the most part, by interested students from my university - mostly psychology students - who were either traveling to foreign countries or returning to their home countries for the summer, and who were anxious to participate in this research. To our good fortune, it has turned out that these „volunteers" were preparing to leave for countries covering most of the world.[8] Clearly, however, the final sample of countries has to some degree been dictated by convenience.

In most countries, data have been collected in either the largest city or a rival major city: Amsterdam (The Netherlands), Athens (Greece), Budapest (Hungary), Dublin (Ireland), Frankfurt (Germany), Guanzhou (China), Hong Kong (Hong Kong),[9] Jakarta (Indonesia), London (England), Mexico City (Mexico), Nairobi (Kenya), New York City (USA), Paris (France), Rio de Janeiro (Brazil), Rome (Italy), San Jose (Costa Rica), San Salvador (El Salvador), Seoul (South Korea), Singapore (Singapore), Stockholm (Sweden), Taipei (Taiwan), Tokyo (Japan), Toronto (Canada) and Vienna (Austria). In four other countries, for various reasons, the observations were made in more than one city. In Poland, data were collected in Wroclaw, Lodz, Poznan, Lublin and Warsaw. In Switzerland, measures were taken in both Bern and Zurich. In Syria and Jordan, most observations were made in the capital cities of Damascus and Amman, but some were done in secondary locations. In each of these cases, data from the different cities were combined for that country. Data for all countries were collected during the summer or other warm weather months of the year, over the period 1992-1994.

Within each city, we have measured the same three indicators of the pace of life as in our original six-country study: First, we timed the walking speeds of pedestrians - in this case over a distance of 60 feet. Second, as an indicator of work speed, we again clocked how long it took postal clerks to fulfill a standard request for stamps. Third, as an

---

8 We are indebted to the following people for their help in collecting data in one or more countries: Karen Bassoni, Gary Brase, Stephen Buggie, Andy Chuang, Holly Clark, John Evans, Robert Lautner, Royce Lee, Andy Levine, Todd Martinez, Michiko Moriyama, Carlos Navarette, Ara Norenzayan, Julie Parravano, Aroldo Rodrigues, David Tan, Jyoto Verma, Karen Villerama and Sachiko Watanabe.

9 For simplicity, Hong Kong is referred to here as a country. Technically, however - for the next couple of years, at least - it is a colony.

estimate of a city's interest in clock time, we again observed the accuracy of bank clocks in downtown areas.

Given that these studies are still in progress, it would be premature to list the country rankings on each measure. However, even at this point, it seems clear that our hypotheses are largely supported by the data (see results in Table 2, p. 133).

Our first group of hypotheses have concerned the prediction of pace of life - specifically, that pace would be related to population size, climate, individualism-collectivism and economic well-being. Three of these predictions were strongly supported. Warmer climates (as measured by average annual high temperatures) were significantly associated with a slower pace of life on all measures. Collectivism (as measured by Triandis' country ratings - described in Diener et al., 1994) showed significant negative correlations to pace on all three measures. Economic vitality was based on three types of measures: (a) gross domestic product per capita; (b) purchasing power parity (an estimate of how much the average income in a country is capable of purchasing); and (c) average caloric intake (an estimate of the ability to fulfill minimal needs).[10]

It may be seen that there are significant positive correlations between economic success and all pace measures for all three of these economic indicators. The single predictor variable which has not consistently related to pace is population size. This is not totally unexpected, however, given that population size has been poorly sampled in this study in that we generally tested only the largest city in each county.

The second group of hypotheses concern the consequences of pace for physical and psychological well-being - specifically, that faster paces would be associated with higher smoking rates, higher death rates from coronary heart disease and higher life satisfaction. It may be seen in Table 2 (p. 133) that two of these three hypotheses have received strong and consistent support. All three pace measures have been associated with both higher rates of coronary heart disease and higher self-ratings of life satisfaction. (The correlations for ratings of life satisfaction are of only borderline significance for two pace measures, which may simply reflect the small sample size on which they are based). Smoking rates have not consistently related to pace: they are, as expected, somewhat higher in places with faster walking speeds, but are virtually unrelated to postal speeds and negatively related to clock accuracy.

---

[10] All statistics are from the World Bank (1994).

	Pace measure		
Community characteristic	walking speed	postal speed	clock deviation
Population size	.05 (26)	-.36* (26)	.01 (26)
Climate (temperature)	-.62** (28)	-.32* (28)	-.56** (28)
Collectivism	-.35* (28)	-.41* (28)	-41* (28)
GDP per capita	.57** (26)	.54** (26)	.57** (26)
Purchasing power parity	.53** (26)	.49** (26)	.57** (26)
Daily caloric intake	.45** (26)	.36* (26)	.43* (26)
Percent smokers	.30 (13)	.09 (13)	-.57* (13)
Coronary heart disease	.44* (23)	.39* (23)	.33* (23)
Life Satisfaction	.40 (15)	.56* (15)	.36 (15)
Helping Behavior			
Blind person	.02 (18)		
Hurt leg	.40* (18)		
Dropped pen	.23 (18)		

*p < .05; **p < .01 using one-tailed significance test.

Table 2: Correlations[11] (and sample sizes)[12] between the pace of life and other selected community characteristics from 28-country study

[11] Higher correlations indicate that the community characteristic is positively related to a faster pace of life (faster walking speeds, faster postal speeds, and smaller clock deviations).

[12] Note that statistics for some community characteristics were not available for some countries resulting in smaller sample sizes for some analyses.

Third, it was predicted that pace would be negatively related to the tendency to help strangers. This hypothesis is being tested in another study still in progress. Basically, we have conducted three helping experiments in a sample of what is currently 18 international cities (many, but not all, of which are the same as those in the pace of life studies under discussion). Briefly, the three helping experiments are: (1) Helping a Blind Person Across the Street. In this experiment the researcher, dressed in dark glasses and carrying a white cane, acts the role of a blind person needing help getting across the street. Just before the light turns green, he steps up to the corner, holds out his cane, and waits for help. A trial is terminated after the lesser of either 60-seconds or when the light turns red. (2) Dropped Pen: walking at a moderate pace, the experimenter approaches a solitary pedestrian passing in the opposite direction. When 15-20 feet away, the experimenter reaches into his pocket and 'accidentally,' without appearing to notice, drops his pen behind him, and continues walking. Helping is scored on a 5-point scale, ranging from no help offered to picking up the pen and running back to hand it to the experimenter. (3) Hurt leg: walking with a heavy limp and wearing a large and clearly visible leg brace, experimenters 'accidentally' drop, and then unsuccessfully struggle to reach down for, a pile of magazines as they come within 20 feet of a passing pedestrian. Helping is scored on an incremental scale ranging from no help offered to picking up the magazines and asking if they could be of further assistance. Finally, our measure of the pace of life in these same cities is walking speed, which is measured in the same manner as in our other experiments. (For more detailed description of the methods used in the helping experiments, see Levine et al., 1994).

It may be seen in Table 2 (p. 133) that the hypothesized relationship between walking speed and helping is partially supported by the data gathered thusfar. As predicted, there appears to be a significant negative relationship between walking speed and the likelihood that assistance will be offered to someone with a hurt leg, and a non-significant negative relationship between walking speed and helping in the dropped pen situation. There appears, however, to be virtually no relationship between walking speed and help in the blind person situation. We believe that the latter may be due to either of two factors: first, the nature of the blind person experiment requires that potential helper is standing still at the critical helping moment (they are waiting at a red light), and so their walking speed may not be as relevant as in the other conditions; and, second, we have found that norms governing assistance to blind people are structured very differently and often explicitly from country to country: In Japan, for example, the changing traffic lights are often accompanied by sounds intended to signal the blind. Thus, we believe that the correlations obtained in the hurt leg and dropped pen situations may be more valid tests - and at least partial confirmations - of the walking speed/helping hypothesis.

Figure 2 (p. 136) summarizes our hypotheses and the major findings obtained at this point in our study. It is proposed that norms about the pace of life are central to a wide range of community characteristics, and that the relationship between pace and these other characteristics is frequently mutually reinforcing. We suggest that factors such as population size, climate and cultural values may raise or lower norms about pace. These norms may, in turn, have an effect on the population size (slower places may seem more attractive to migrants) and cultural values (e.g., as explained above, individualism may be a result of faster norms). Although not tested in this study, it is also suggested that migration patterns will be selective - people attracted to faster places may be the very ones who are likely to support a fast pace of life, and vice versa. Faster paced places will tend to be more economically productive (which, in turn, will raise the value of time and, subsequently, the pace of life).

The consequences of economic vitality should, we predict, be mixed. On the one hand, the very time stressors that are responsible for such success will be reflected in unhealthy behaviors (e.g. smoking) and stress-related physical difficulties (e.g. coronary heart disease). On the other hand, the tangible rewards of economic success should raise the general level of psychological well-being, as reflected in self-ratings of life satisfaction. Finally, if the *budi bahasa* and Milgram are correct, fast-paced norms and emphasis on economic productivity should result in a sacrifice in attention to the needs of others, as reflected in the amount of help offered to strangers.

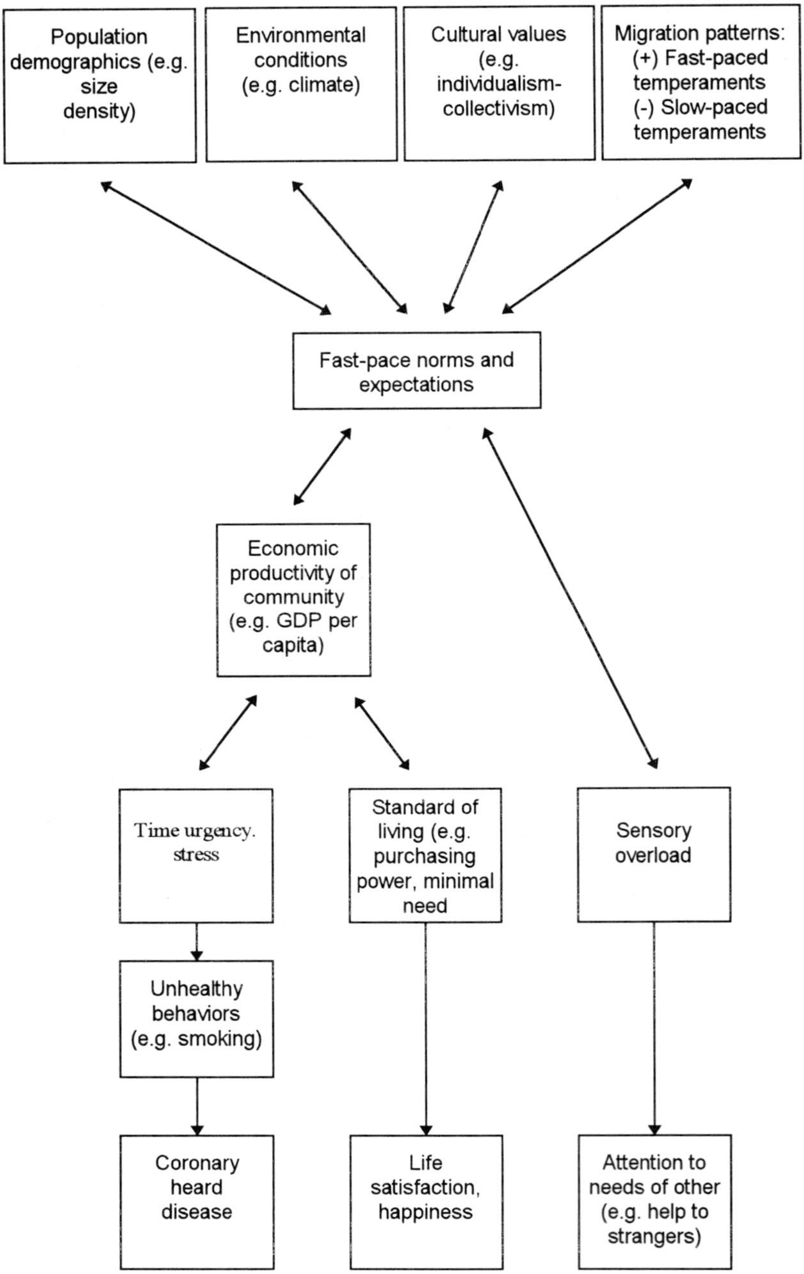

Figure 2: A community's pace of life: some predictors and consequences

Concluding comments

Our studies demonstrate the centrality and significance of the pace of life in the personality of a culture. Our results indicate that it is hardly an exaggeration to describe the pace of life truly as the heartbeat of a culture.

There is evidence, in fact, that the pace of life is the single most significant defining element of social time. A study by Dapkus (1985) demonstrated how the pace of life permeates the entirety of the experience of social time. Dapkus was interested in what categories people use to describe the way they experience time. She conducted a series of interviews with adults who were asked questions about the entirety of their temporal experience, ranging from when they first became aware of time to how they experience time in different situations, to how it has changed over the course of their lives. Their descriptions, Dapkus found, could be classified into three categories: (1) change and continuity, or „becoming in time" - this refers to the understanding that life is a process of change; (2) limits and choices, or „doing in time"- this is the realization that time is limited, thus forcing us to make choices about how we use our time; and (3) tempo, or „pacing in time"- what Dapkus and her subjects describe as „the speed or pattern of movement in time, whether fast or slow, regular or irregular“ - in other words, the pace of life.

Perhaps the most significant finding in Dapkus' study was the extent to which tempo turned out to be intertwined with other statements about time. Tempo, she found, was almost always described in terms of the two other categories: people described tempo by applying it to „change and continuity“ (for example: „As you get older, people say that time appears to go faster;“ or „When my boy lives one extra year, that's 10% of his life, but when I live one extra year that's only 2% of my life.“). People use tempo to describe „limits and choices" (for example: „My husband doesn't feel as rushed as I do; he's more relaxed, he takes it in stride if time runs out. He can say, 'That's that,' but I'd be trying to cram in one more thing.“). They also apply it to both other categories at once (for example: „I still tend to think that I have many more years, which is probably why I'm not more motivated to make changes now. It's easy to think, well, I'll do that after the kids are in college. A person diagnosed as having a terminal illness will quickly set some new priorities and make some changes. Unless you're hit over the head, it's easy to keep doing it the same old way.").

In other words, people do not separate tempo out as a single, isolated feature of their temporal experience. It is both an overall umbrella and an inter-connected part of all of their experience of time. In music attributes like tempo and rhythm are distinct entities operating independently. In the world of social time, however, the lines are less clear cut. It is this broader conceptualization of time that represents what people experience as the pace of life.

It is no wonder that our studies have revealed such strong correlations between the pace of life and other characteristics of communities. The pace of life is the defining element of the experience of social time which, as Oswald Spenger observed, is above all other things what differentiates one culture from another. Our studies have demonstrated that the socio/environmental/economic make-up of a culture is inextricably connected to its pace of life; and the consequences of this tempo are reflected as the very quality of human life.

It is hardly surprising that the returning Peace Corps volunteers described at the beginning of this chapter focused so much of their culture shock on the alien pace of life. The pace of life is central to the social fabric of a culture and, as such, is sensitive to most every cultural nuance. The pace of different cultures varies dramatically and is complex. The pace of life differs from situation to situation, in its cadence, rhythms, sequencing and, perhaps most important of all, in its meaning and significance to those whose lives they regulate. To make matters more difficult for strangers, a culture's rules of time are rarely made explicit. They lie hidden and, as such, are mostly outside of voluntary control - either by cultural insiders, who simply assume the norms to be normal and natural, or by unsuspecting outsiders, who discover how rigidly the norms are enforced. Cultural norms for appropriate tempos in work, in play, in what constitutes punctuality and lateness, and in the meaning of 'now' and 'later' are all aspects of what Hall (1959) describes as the „silent language." There are no dictionaries to translate a culture's temporal rules. These informal patterns of time „are seldom, if ever, made explicit. They exist in the air around us. They are either familiar and comfortable, or unfamiliar and wrong (Hall, 1959, p. 156)." With no formal definitions to help us along, it is little wonder that this silent language can pose as many difficulties for cultural adaptation as differences in the formal spoken language itself.

References

Amato, P.R. (1983). The effects of urbanization on interpersonal behavior. Journal of Cross-Cultural Psychology, 14, 353-367.

Booth-Kewley, S. & Friedman, H. (1987). Psychological predictors of heart disease: A quantitative review. Psychological Bulletin, 101, 343-362.

Bornstein, M.H. (1979). The pace of life: revisited. *International Journal of Psychology, 14*, 83-90.

Bornstein, M. & Bornstein, H. (1976). The pace of life. *Nature, 259*, 557-559.

Cohen, J.B., Syme, S.L., Jenkins, C.D., Kagan, A., & Zyzanski, S.J. (1975). The cultural context of type A behavior and the risk of CHD. *American Journal of Epidemiology, 102*, 434.

Dapkus, M. (1985). A thematic analysis of the experience of time. *Journal of Personality and Social Psychology, 49*, 408-419.

Diener, E., Diener, M. & Diener, C. (1994). *Factors predicting the subjective well-being of nations*. Unpublished manuscript.

Fiske, A.P. (1992). The four elementary forms of sociality: framework for a unified theory of social relations. *Psychological Review, 99*, 689-723.

Friedman, A.P., & Rosenman, R.H. (1974). *Type A behavior and your heart*. New York: Knopf.

Hall, E.T. (1959). *The silent language*. New York: Doubleday.

Hoch, I. (1976). City size effects, trends and policies. *Science, 193*, 856-863.

Hoel, L.A. (1968). Pedestrian travel rates in central business districts. *Traffic Eng., 38*, 10-13.

Howard, C.G. (1980). The expatriate manager and the role of the MNC. *So. Personnel Journal, 59*, 838-844.

Levine, R. (1988). The pace of life across cultures. In J. E. McGrath (Ed.), *The social psychology of time: new perspectives* (pp. 39-62). Newbury Park, CA: Sage.

Levine, R. (1990, September-October). The pace of life and coronary heart disease. *American Scientist, 78*, 450-459.

Levine, R., & Bartlett, K. (1984). Pace of life, punctuality and coronary heart disease in six countries. *Journal of Cross-Cultural Psychology, 15*, 233-255.

Levine, R., Lynch, K., Miyake, K., & Lucia, M. (1989). The type A city: coronary heart disease and the pace of life. *Journal of Behavioral Medicine, 12*, 509-524.

Levine, R., Martinez, T., Brase, G. & Sorenson, K. Helping in 36 U.S. cities (1994). *Journal of Personality and Social Psychology, 67*, 69-81.

McGrath, J.E. (1989). The place of time in social psychology: some steps toward a social psychological theory of time. Paper presented at the Seventh Conference of the International Society for the Study of Time, July, 1989, Glacier Park, Montana.

Marmot, M.G., & Syme, S.L. (1976). Acculturation and coronary heart disease in Japanese-Americans. *American Journal of Epidemiology, 104*, 225-247.

Matthews, K. (1988). Coronary heart disease and type A behaviors: update on and alternative to the Booth-Kewley and Friedman (1987) quantitative review. *Psychological Bulletin, 104*, 373- 380.

Milgram, S. (1970). The experience of living in cities. *Science, 167*, 1461-1468.

Raybeck, Douglas (1992). The coconut-shell clock: time and cultural identity. *Time and Society, 1(3)*, 323-340.

Smith, T., & Anderson, N. (1986). Models of personality and disease: An interactional approach to type A behavior and cardiovascular risk. *Journal of Personality and Social Psychology, 50*, 1166-1173.

Spradley, J.P. & Phillips, M. (1972). Culture and stress: a quantitative analysis. *American Anthropologist, 74*, 518-529.

Triandis, H. (1994). *Culture and social behavior*. New York: McGraw-Hill.

Wright, L. (1988). The type A behavior pattern and coronary artery disease. *American Psychologist, 43(1)*, 2-14.

U.S. Department of Health and Human Services (1987). Regional variation in smoking prevalence and cessation: Behavioral risk factor surveillance, 1986. *Morbidity and Mortality Weekly Report, 36*, 751-754.

World Bank (1994). *The World Bank Atlas: 1995*. Washington, D.C.: World Bank.

Part V:

Toward a unified model of psychological time

Chapter 8: Are we coming near a general model of psychological time?

Some questions from the comparative and developmental approaches*

HELGA LEJEUNE and MARC RICHELLE

Introduction: how distinct are humans?

Chronobiological studies have abundantly shown that humans share with other animals fundamental properties of basic biological rhythms. If civilization has artificially modified a number of natural factors, it has not really liberated humans from the constraints of long established cycles, especially the circadian periodicity. Modern societies, by pushing to an extreme degree the changes in time schedules, be it for work or leisure, have indirectly revealed how strongly our species depends upon biological time constraints. Do we similarly share with other animals major properties of psychological time? This is the question this chapter addresses. The answers which shall be proposed are incomplete and far from being decisive. Reasons for the situation are, among others, that our knowledge on psychological time from a comparative point of view is much less developed than is the case for biological time, and, more importantly, drastic differences between lower animals and humans are likely to derive, as in many other aspects of behavior, from the use of language in our species, with all its consequences on psychological functions.

As pointed out in the introductory chapter, the study of psychological time has been stagnant for many years, attracting only a few experi-

* Preparation of this contribution has been supported by Project Grant n° 31 from the Belgian Federal Ministry of Scientific Policy, within the P.A.I. Programme 1990-1995.

menters, working essentially on human subjects, because appropriate techniques for parallel studies in animals were lacking. Progresses in experimental procedures, mainly in the operant conditioning laboratory, provided the experimenters, since the fifties, with efficient tools to explore psychological time in animals. Abundant empirical data were collected and theoretical models designed to account for them (for a review of the work carried out until the late seventies, see Richelle & Lejeune, 1980). Interestingly enough, thanks to operant techniques, the quantity of data collected on animals quickly surpassed the data available on humans (traditionally obtained in the frame of psychophysical methods). One reason for this success was that human subjects cannot help resorting to mental chronometric tricks when submitted to time estimation, reproduction or production tasks involving durations beyond the range of a few hundreds milliseconds. Such contaminating variables escape the experimenter's control. A second reason is that some aspects of temporal regulations of behavior appear only after prolonged exposure to experimental conditions, which is difficult to require from humans.

There is little doubt that the renewal of psychologists' interest in time, and the impressive accomplishments in the last 30 years are at least partly due to contributions from animal studies. These studies, besides the enormous amount of data already mentioned, have generated theoretical models of behavioral and cognitive time which have been taken over or adapted by investigators working on psychological time in humans (for example, see Block & Zakay, this volume, chapter 9). The time is ripe, not for a general theory of psychological time, but at least for local synthesis from a comparative approach. A general theory which eventually will lead us to identify cross-species commonalties, which humans share with other animals, as well as specificities linked with species-specific features (humans possibly having in this respect a unique position because of their specific cognitive and/or linguistic endowment). As in other fields of psychology and sociology, the issue of human specificity is doomed to contamination by unscientific presuppositions, and, depending on one's own inclination, emphasis will be put on continuities or on discontinuities. Some have argued that the capacity to estimate durations, be it of external events or of one's own behavior, does not basically change from lower animals (or at least from lower vertebrates) to man, and have hypothesized that psychological duration is the foundation from which eventually derive the most complex elaborations such as time concepts, temporal reasoning, experience of time and the like, possibly specific to humans (for a defense of continuity see Richelle, 1968, Richelle et al., 1985). On the opposite side, others argue that duration itself, whatever its phylogenetic history, is in humans an elaborate product of cognition (see, for example, Michon, 1990). In the middle position are those researchers who admit that the perception of duration is very much the same in humans and in animals, while other aspects of psychological time (conceptual or subjective) are typically human (for a brief discussion of the issue, see Richelle, 1992).

Methodological caveat

Although the following is rather trivial from a methodological point of view, some words of warning are called for concerning the type of comparison(s) from which conclusions can be drawn. A sound comparative approach would require careful study of a number of species, both very distant and very close phylogenetically, so that the status of differences and resemblances observed can be adequately interpreted. However, most discussions are still based on the study of a very limited number of species. As could be expected, most researches on psychological time have been carried out on traditional laboratory species, i.e. rats, pigeons, occasionally monkeys. The anthropocentric trend of the dominant cognitivist school of thought has not favored the exploration of a wider range of species, the results of which might question the validity of theoretical models proposed. We shall come back to this issue later.

Similarly, models are often built upon and validated by empirical evidence gathered in one or a few selected experimental situations, disregarding the diversity and the complexity of temporal regulations when observed in a variety of contexts and constraints. The rationale for giving privilege to one procedure is quite weak, and consequently the generality of any model so constructed is weak itself. We shall turn in a moment to the most popular model in the field, but one should keep in mind that it is essentially based on data collected in animals using the so-called peak procedure, undoubtedly an interesting response timing technique, but far from condensing all what would appear relevant to an understanding of psychological time.

Leaving out Pavlovian and a few other procedures, operant conditioning offers a wide variety of schedules to explore temporal regulations. A complete description of them would take more space than available here (it can be found in Richelle & Lejeune, 1980; Lejeune, 1993; Lejeune & Macar, 1994). Suffice it, to remember the main categories of these procedures, not for the sake of showing the imaginative production of experimenters, but to point to the crucial problems they raise to a theory of psychological time. If a model, such as Church's model that will be discussed below, draws its empirical evidence mostly from one temporal regulation procedure, such hyperselective choice must be justified by demonstrating that other procedures do not add only marginal or irrelevant information on psychological time. Failing to provide such demonstration, it must account for any empirical data as provided by any other procedure.

A major distinction must be drawn among operant procedures between those which bear upon a subject's capacity to estimate the duration of external events and those which permit an analysis of a subject's capacity to control adequately the occurrence of its own activities in time. In the first case, time is processed as any sensory information (in spite of the fact that no receptor is specifically devoted to time as a stimulus), while in the second case, time is integrated in the regulation of action. Time perception is complemented by timing. Some procedures are especially appropriate for the study of the first aspect, for example by using discrimination tasks in which the subject has to tell the difference between an external stimulus of a given duration and a standard stimulus. Other procedures, such as the *Fixed-Interval schedule* (FI) or the schedule of *Differential Reinforcement of Low response rates* (DRL), are designed to study the timing of motor responses. Do these two categories of temporal regulations proceed from common mechanism(s) and obey the same laws? Or do they reflect different processes? The question has generally not been addressed because cognitive psychology has placed more emphasis on information processing, at the expenses of the motor, efferent side of behavior.

Another no less important distinction among procedures is related to what could be called the adaptive value of temporal regulations. In short, we are confronted here with a paradox. In some procedures, such as the FI schedule, temporal regulation appears spontaneously, entrained, so to speak, by the periodicity of the reinforcement (or strictly speaking of the availability of the reinforcer). In that situations, subjects develop very typical regular patterns of responding, exhibiting long pauses after reinforcement, followed by anticipatory activity as the time of reinforcement approaches, while „waiting" is by no means a condition for reinforcement. In other procedures, like DRL, the response is to be postponed until a critical duration has elapsed, usually since the previous response, making the timing of behavior the very condition for reward (such procedures are also known as temporal differentiation procedures, because they select or differentiate a temporal characteristic of behavior). The temporal regulation can be said to be spontaneous in the first case and required in the second. Paradoxically, the same subject is able to withhold responding for much longer periods of time in the first case, when it is not necessary, than in the second, when it is required.

Finally, exploring different procedures is also intended to focus more specifically on the timing processes proper, in other words, to capture psychological time in its purest manifestations, getting rid or controlling all sorts of presumably irrelevant variables, such as motivational state, attentional constraints, species-specific limitations, etc. For example, rather than using short operant responses as is usually the case in DRL schedules, which leaves the subject's activities between responses uncontrolled, one can use a prolonged response, in which the subject must maintain for a critical duration if it is to be rewarded (*Differential Reinforcement of Response Duration schedule*, DRRD). One variant of the FI schedule, aimed at isolating the timing mechanisms proper, has been popularized by Roberts (1981) and Church (1984) after an earlier experiment by Catania (1970) as the *peak procedure*. Besides psychophysical procedures (bisection and generalization), it has provided an important amount of data used to support Church's model.

Formal models

Currently discussed formal models of psychological time in humans are mainly derived from psychophysical studies (see Eisler, this volume, chapter 4). They are based on empirical data collected in experiments on the discrimination of duration (differential threshold) and on scaling (the bisection procedure being the most widely used). Without entering the intricacies of a highly sophisticated field, let us mention that on the whole psychophysical studies both on animal and humans confirm that Weber's Law applies to time as it does to other stimulus dimensions. In other respects, some motor performances are best described by using a linear function ($M = b\,t + a$), others by a power function ($M = k\,t^{n}$) [where M is the average performance and t the temporal parameter of the task; a, b and k are fitted constants].

As already alluded to, the most popular model is the *scalar model* (see also Block & Zakay, this volume, chapter 9). It was initially proposed by Gibbon (1977), but had been in some respects anticipated by Treisman (1963). However, it is better known as Church's model, having been especially elaborated, and subsequently modified according to empirical and theoretical progresses or shift of emphasis, by Church and his co-workers (Church, 1984; 1989; Church & Broadbent, 1990; Gibbon & Church, 1990; 1992; Church et al., 1994). It is essentially a description in terms of formal variables, to which psychological variables are matched. Both the description in formal and in psychological terms have evolved with time, paralleling the evolution of ideas in cognitive psychology.

In its classical form, which could be labeled „temporal information processing model", the model involves, as shown in Figure 1 (p. 149) three levels, i.e., a timer proper, memories and a decision unit. At the lower level, an internal time base sends pulses, via a switch, to an accumulator. The latter - at the second level - feeds the working memory if there is a delay between the duration to be measured and the opportunity to respond (as is the case in *symbolic matching-to-sample* or *duration discrimination procedure*). The content of the accumulator or of working memory is permanently being assessed with respect to the reference memory in which relevant information as to the duration to be estimated are stored. At the third level, decision is made after a comparator matches the information processed in working memory to those stored in the reference memory. Component boxes can be viewed as plausible inferences as to the processes underlying time estimations, although some of them could be questioned and revised - in fact have been questioned and revised (see Church & Broadbent, 1990) - along the connectionist line of thought. The scalar model has essentially two distinctive features deriving from its very structure. Especially central for the model is the place of the reference memory which provides the standard for comparison before each decision to respond occurs: (1) the reference memory stores the estimations of the critical duration in the shape of a Gaussian distribution, with the variance being a function of the square of the mean; (2) mean and standard deviation of this distribution are proportional to the critical duration. In other words, the performance should exhibit two characteristics if the model is to be validated: (1) the central tendency of time estimations should coincide with the critical duration to be estimated and (2) the coefficient of variation (that is the ratio of the standard deviation to the mean, s/m), the major evidence of an organism's sensitivity to time, should remain constant whatever the absolute value of the critical duration. If the latter is modified, the subject will „re-scale" its timing units according to Weber's Law.

The scalar model, initially elaborated in the context of animal studies, has been applied to human data as well (Wearden, 1991a). Several studies have shown its adequacy to account for a number of empirical data, suggesting that it could be seen as a robust candidate for a general model, especially in unifying human and animal time. However, even if particular samples of behavior do fit the assumptions of the model, the question remains open as to its generality, *within* and *across procedures*, *within* and *across species*, and throughout *development*.

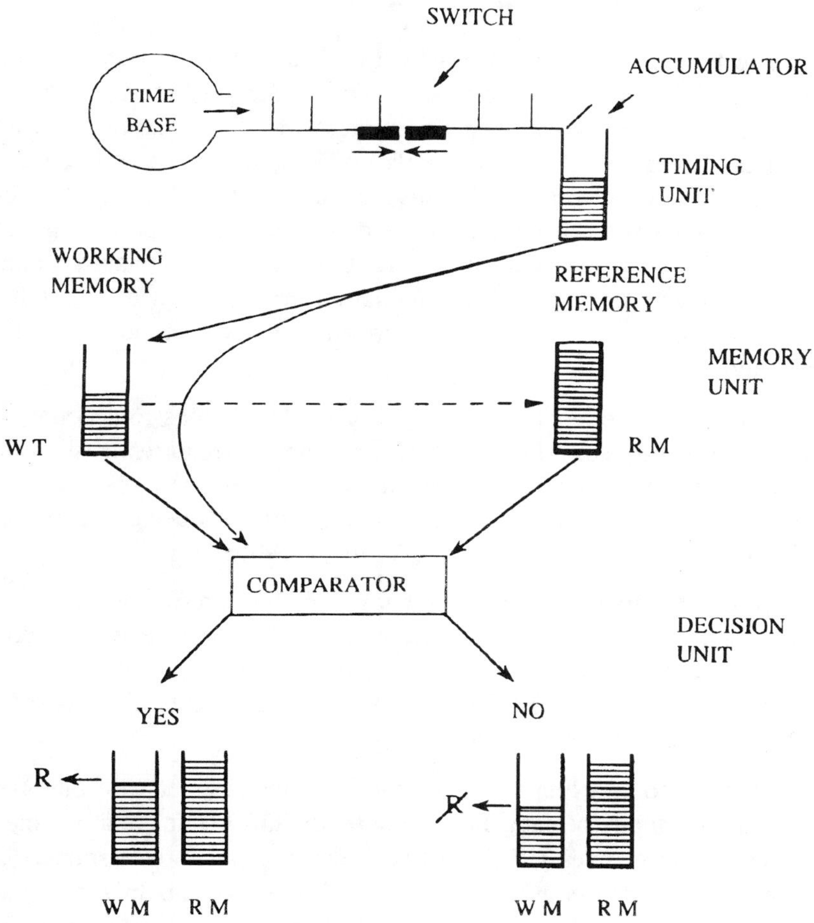

Figure 1: The internal clock model for processing of temporal information (after Church, 1984)[1]

[1] The six components of the model are organized in three levels (counter, memory, decision). When confronted to a duration to be estimated, the time base produces pulses which are sent through a switch to the accumulator where they are counted (these pulses are transferred to a working memory if there is a gap between the end of the duration to be estimated and the opportunity to respond, as is the case with duration discrimination procedures). The comparator compares the duration in the current trial with a representation of the duration drawn from the reference memory, where durations reinforced during previous trials are stored. The decision to respond (*yes*) or not (*no*) is taken according to a norm of similarity between these two values.

In search of pure time

The first point has been dealt with in several now classic reports (Catania, 1970; Platt, 1979; Stubbs, 1979; Zeiler, 1986). According to Zeiler (1983; 1986), Weber's law does not always fit temporal differentiation data because behavior is jointly governed by timing and non-timing processes which cannot be disentangled. „Pure timing“ (Zeiler, 1985) is mixed with behavioral noise depending upon non-timing factors. Despite such a pessimistic view, the quest for unbiased timing within the frame of temporal differentiation or regulation data has followed five main roads which are described below with reference to the scalar model.

In the first, the model is corrected or modified so as to account for discrepancies between observed behavior and the requirements of the model. These discrepancies are seen as the result of the overlapping of two processes, a random response production process being superposed to the scalar timing process proper (Wearden, 1985).

Secondly, attempts have been made to design interference-free procedures suited to unveil pure clock time, such as the peak procedure where, at least theoretically, response rate and response timing are truly dissociated (Catania, 1970; Roberts, 1981; an earlier attempt was made by Mechner & Guevrekian, 1962).

Thirdly, some experimenters have been looking for rare bias-free cases among empirical data. For example, the differential reinforcement of inter-response times (IRTs) between discrete perching responses in pigeons yielded data in agreement with both requirements of scalar timing theory. The central tendency of the IRT distributions matched the reinforcement criterion and the coefficient of variation remained approximately constant over durations ranging from 10 to 70 seconds (Jasselette et al., 1990).

Fourthly, obstacles to the capture of pure time might depend on interindividual variations. For instance, only those subjects were retained for peak procedure training (18 out of 36) which exhibited the highest response rates on the second of a 2-session random-interval pretraining (Roberts & Holder, 1984, experiment 1).

Finally, obstacles to the capture of pure time have been attributed to intraindividual variations. Consequently, parts of the data obtained on individual subjects have been discarded, because they did not fit the assumptions of the model. This now widely used strategy has progressively drifted from a „reasonable“ grooming of the data (discarding only extreme „deviant“ response patterns) towards dropping high proportions of the performances of individual subjects. Recent cases are for example described in papers discussing multiple timing or the sources of variance in scalar timing, on the basis of single trial analysis. For instance, are retained individual break-run-break patterns characterized by „good starts" and „good stops", the criterion „good" being defined with regard to the critical temporal parameter, and thus to the expected response timing (Cheng et al., 1993; Church et al., 1994; Leak & Gibbon, 1995). What one would expect, of course, from such a strategy, is a validation of the model. Ironically, as the search goes on, and as increasingly sophisticated procedures are put to work in collecting or analyzing data, pure timing seems to remain elusive.

In the first solution, appeal has been made to various „extra-temporal" variables, such as motivation and attention. Attention has been especially in favor in the recent years. However, the idea is not new. Attentional effects were discussed in the frame of the initial version of the temporal information model, in which the different stages of time processing were phrased in psychological terms. Attention was considered as a peripheral factor acting either at the level of the switch or when a sample duration is retrieved from reference memory. Indeed, experiments using the prior entry or the unbalanced probability methods in bisection or peak procedure showed that attentional factors can modulate performance (Meck, 1984). In attentional models of timing (for example, Zakay, 1989), the limited attentional resources are supposed to be distributed between a timer (analogous to the clock stage in Church's model) and a processor devoted to the treatment of non-temporal aspects of stimuli. If, in double task procedures, attention is directed towards the stimulus processor, errors in time estimation will increase, and vice-versa (Macar et al., 1994). So far, the attentional model has been thoroughly tested with human subjects only. If a parametrical relationship between attention and timing could be drawn from analogous experiments run with animal species, the part of behavior neglected in solution 5 might be rehabilitated, eventually resulting in a more comprehensive account of performance. In this respect it is worth noticing that most recent theoretical attempts integrate the information processing and attentional models (see Block & Zakay, this volume, chapter 9).

Attention is, however, only one of a number of variables plausibly affecting performance. Inhibition might be another one. The hypothesis of a balance between the ability to inhibit motor output (and thus operant responses) and the quality of response timing was proposed by Richelle (1972), not to mention Pavlovian conditioning theory much earlier. This hypothesis has so far remained dormant and still awaits experimental tests, tests admittedly more difficult to design than in the case of attention, because inhibition is an internal variable difficult to control as a independent variable (the only indirect attempts are the classical Pavlovian disinhibition tests, and a few operant analogues; see Richelle & Lejeune, 1980).

The last two solutions raise basic problems to which psychologists have been confronted on many occasions: how should we deal with between and within subjects variations? Can we legitimate a selection of subjects among a sample, or of data among the whole range of observation in individual subjects? Neutralizing between-subjects variability by analyzing only group results has been another, more conformist way to avoid the difficulty. Referring again to the peak procedure, response rate functions have usually been derived from averaging group performance over several sessions. To what extent are such functions really representative of time estimation capacities in the species is by no means clear. Close examination of individual performances would suggest that it is not the case. For instance, average data collected on senescent rats using the peak procedure (Lejeune & Soffié, in preparation) seems to confirm response rate functions described by Meck et al. (1986), that is, the peak time is shifted to the right. However, inspection of individual response rate functions revealed that three out of 10 older subjects still matched young adults, whereas the rightward shift was observed in the others.

The fifth solution, i.e., rejecting some of the individual trials as „bad," might, at first sight, be legitimated by assuming a two-processes model, as considered in solution one above. But assuming that pure time is contaminated by some other factors does not dispense with need to identify the factors at work, and does not legitimate a priori rejection of some data; since rejection could only be acceptable if based on explicit criterion, which in turn implies that contaminating factors have been identified. Experimenters in the 60's and 70's were confronted with the problem, but did not avoid it by a priori rejection. Early data collected with the DRL procedure provided evidence for two distinct types of IRTs, some of which very short and limited by responses emitted in „bursts," others were much longer, matching the critical delay for reinforcement. These bursts were systematically studied and eventually fairly well explained, and legitimately distinguished from timing responses proper (for a review of these classic studies, see Richelle & Lejeune, 1980).

The comparative data

Cross-species comparisons are obviously a prerequisite to any general theory of psychological time, as well as to any specific theory of human psychological time. Such research strategy has been shown to be productive in the field of chronobiology, unveiling both universal and species-specific characters of biological rhythms. Curiously, it has hardly been applied to the study of time in behavior.

Most animal data have been collected on traditional laboratory species, i.e., pigeons and rats, occasionally monkeys. Very few studies have been carried out with other species. Richelle & Lejeune (1980, 1984) mention about thirty studies, some of which anecdotal, and therefore of no use in the validation of models such as Church's model. Since that time, a few more species have been systematically studied, in a comparative perspective. Still, the limited interest in that approach is striking. Whenever animal studies are invoked in the field of psychological time, it is more with the concern for generalizing processes supposedly at work in humans than with the goal of understanding throughout phylogenetic history the organisms' capacities to cope with time, eventually up to the complex processing of time emerging in humans - if it can be demonstrated that it is the case (see Wasserman, 1993; Wearden & Lejeune, 1993).

Besides the numerous studies on rats, pigeons and, to a lesser extent, monkeys, empirical data are available on domestic cats (*Felis cattus*), various strains of mice (C27/black, NMRI, Balb/C among others), field-mice (*Apodemus sylvaticus*), gerbils (*Meriones unguiculatus*), quails (*Coturnix coturnix japonica*), cichlid fishes (*Tilapia nilotica L.*), fresh-water turtles (*Pseudemys scripta elegans W.*) (Lejeune, 1971; 1993; Lejeune & Nagy, 1986; Lejeune & Wearden, 1991). The sample is too small to permit conclusions as to universals of time which all species would share. In fact, a crude analysis of performances across species shows that some of them are more skilled than others in their temporal regulations. Mammals and birds, for example, perform better than fishes or turtles. At first blush, this would suggest that the mechanisms of the hypothetical internal clock, or for that matter, of the scalar model, cannot be reduced to any simplified picture.

Closer analysis of available data, however, brings some support to the generalizability of the model. The coefficient of variation of temporal estimators as produced by the subjects indicate that, at least in some cases, sensitivity to time remains constant along different values of the critical delay; this is one of the requirements of the scalar model, as will be discussed later. The second requirement, that is the matching of the central tendency of temporal estimators to the duration to be estimated, is also fairly well met, at least in some species. Thus, in spite of cross-species differences in their sensitivity to time, reflected in the absolute value of the coefficient of variation, the constancy of the coefficient in a given species points to one important common feature. Figure 2 (p. 155) shows this phenomenon, and suggests that the scalar model could be compatible with different levels of sensitivity to time in different species, possibly derived from selective pressure in the peculiar environment in which they have evolved. In other words, the mechanism of the internal clock could be the same, and the differences observed between species are due to differences in the temporal resolution of the clock (just as chronometers can have different level of accuracy, to one tenth, one hundredth, or one thousandth of a second).

While coefficients of variation obtained under certain conditions remain constant, exceptions occur, in which the coefficient increases, or to the contrary decreases, as the time criterion increases. Data collected on fresh water turtles and cats illustrate the first case; turtle doves provide an example of the second case (Figure 3, p. 157). No explanation has been proposed for the decrease of the coefficient of variation. The increase could result from the combination of a scalar process, the sensitivity of which remains constant, with a random response generating process, as hypothesized by Wearden (1985); balancing adequately the two components in a computer simulation has provided data matching those observed in real performances (Lejeune & Wearden, 1991).

The results discussed above would, at first sight, seem compatible with an evolutionary hypothesis, in which capacities for temporal regulations would correlate, for example, with an index of complexity of the central nervous system throughout the phyletic tree. However, the picture becomes more complicated as one looks at the performances in zoologically close species: important differences are observed, for example, between turtle doves and homing pigeons, or between laboratory mice of various strains and their wild cousin, the field-mouse (Lejeune, 1993).

The student of time enters further puzzlement when confronted with differences in timing capacities within the same species - within the same individuals of a given species - depending upon the sensory modality or the type of response involved in the experimental situation. Rats have been shown to be better „time estimators" when an auditory stimulus is used rather than a visual one, while the converse relation holds in pigeons (Figure 4, p. 157; Meck & Church, 1982; Roberts et al., 1989).

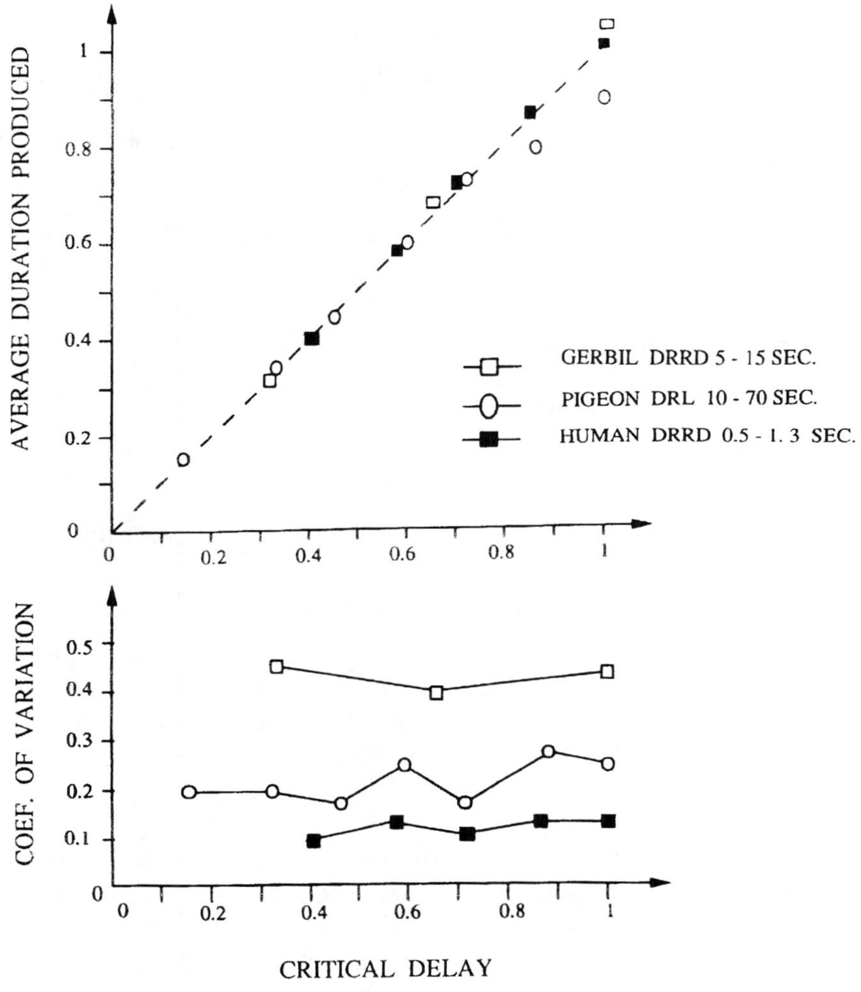

Figure 2: Average duration produced (above) and coefficient of variation of the time estimators (bottom) in time production tasks. Abscissa: value of the critical delay to reinforcement, expressed as the fraction of the maximum delay used in gerbils (empty squares), pigeon (circles) and human adults (filled squares). (After Lejeune & Richelle, 1990, and Wearden, 1991a)

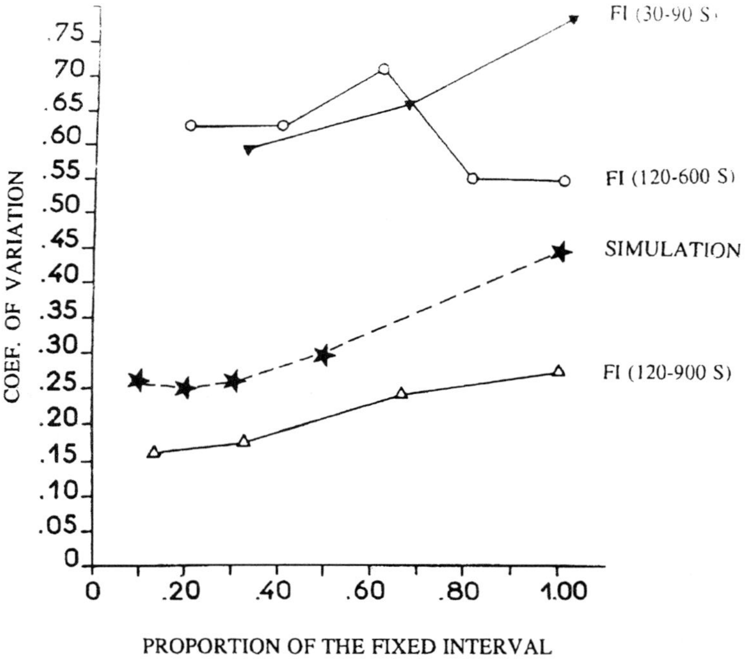

Figure 3: Evolution of the coefficient of variation in temporal regulation under Fixed-Interval (FI) schedules of reinforcement as a function of the interval duration.[2]

Similar conclusions are drawn from studies on cross-modal transfer of time estimations, from studies in which the probability of signals addressing different modalities has been manipulated and from studies exploring parameters such as intensity or „empty" versus „filled" stimulus duration (Meck & Church, 1982; Meck, 1984; Wilkie, 1987; Mantanus, 1981). These effects, however, seem dependent upon the length of training, and fade out after long exposure to time contingencies (Meck,

[2] Abscissa: proportion of the interval used in fresh-water turtles (filled triangles), turtle doves (circles) and cats (empty triangles). The range of durations explored in each species is mentioned at the right of the corresponding curve. Stars joined by broken line show values obtained from computer simulation, applying a complex model which combines a timing process and a nontemporal process in response production (the coefficient of variation of the timing process proper being 0.20, and the probability for a response not controlled by time being 0.05). For simulated data, FI intervals (abscissa) correspond to arbitrary units (after Lejeune & Wearden, 1991).

1984; Roberts et al., 1989), a possible condition to have access to „pure time“ - if it is accessible at all. Indeed, rats trained on an „around the clock“ FI 60 schedule where 30 minutes conditioning sessions alternated with 90 minutes „rest“ periods displayed after 25 days (i.e. 290 training sessions) low response rates and post-reinforcement pauses much longer than those usually obtained with the classical one-session-a-day practice and a smaller number of training sessions. In such a case, distributions of postreinforcement pauses look like a response rate function in peak procedure, with the average post-reinforcement pause close to 60 seconds (Richelle et al. 1985; Lejeune & Richelle, 1990).

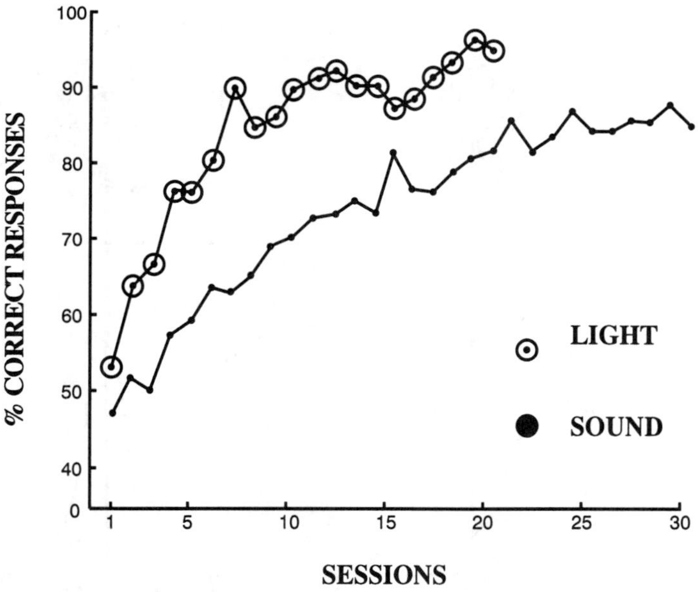

Figure 4: Acquisition curves of pigeons in a time discrimination task involving either visual (white circles) or auditory (black circles) stimuli lasting 2 versus 8 seconds (after Roberts et al., 1989)

Species-specific constraints appear even more obvious when performances are compared in the same species using different responses. A striking example has been described in pigeons, which performed much better on a DRL schedule when a perching response was required than when the traditional key-peck was used (Lejeune & Richelle, 1982; Figure 5, p. 158). Moreover, the coefficient of variation was fairly constant for increasing values of the critical delay, meeting the requirement of the scalar model (Jasselette et al., 1990). That study and others on response topography (for a review, see Lejeune, 1993) indicate that not all motor responses are equivalent in reflecting timing capacities of a given species, one major factor being the response amenability to central inhibitory control, as suggested by refined analysis of key pecks in pigeons exposed to DRL contingencies. „Bursts" of responses emitted with very short inter-response times (IRTs) do not have the status of real operant, and could be interpreted as some sort of displacement activity compensating for the inhibitory tension hypothetically induced by any time contingencies (Richelle, 1972). The fact that they have very little influence on the optimization of performance, as shown by Wearden (1991b), supports that interpretation.

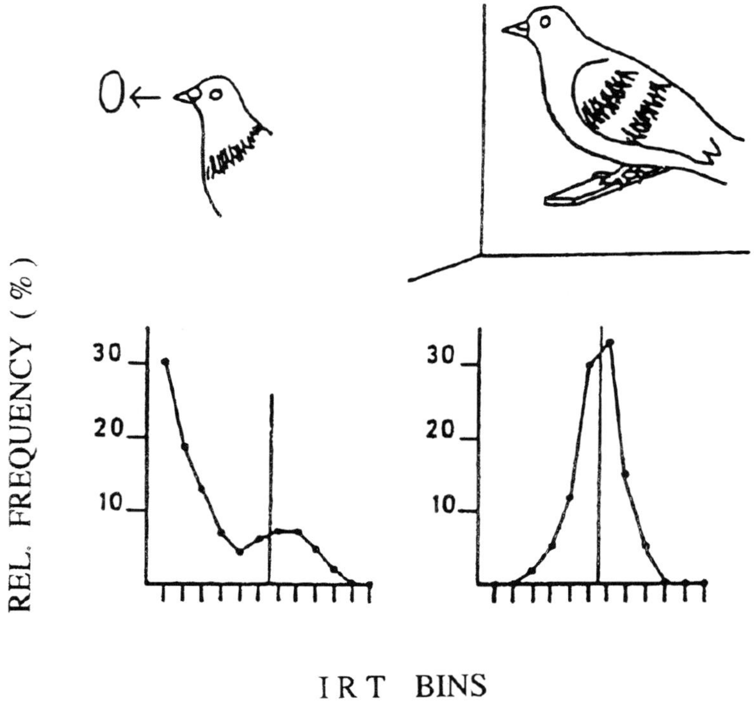

Figure 5: Inter-response times (IRT) curves obtained in a pigeon in a DRL schedule of reinforcement with a key-peck response (left) and with a perching response (right). Ordinate: relative frequencies; abscissa: IRT bins (after Lejeune, 1989a)

Empirical data collected on human adults using operant procedures, usually covering a very limited range of durations, have been compared with animal data, especially in attempts to validate the scalar model

(Wearden, 1991a). They converge with data obtained with other methods traditionally used in psychophysics, which have led to models quite close to the scalar model (Allan, 1992). This, however, leaves out some interesting observations made on humans in some operant procedures such as the Fixed-Interval schedule. It has been shown that, as a rule, human adults do no exhibit the pattern of behavior usually observed in animals (which gave the old cumulative curve its typical scalloped shape, as shown in Figure 6, p. 160). Human subjects either pause for a very long time (approximating the critical interval) after each reinforcement, ending up with one or very few responses sufficient to bring the reinforcer, or they produce a sustained train of responses throughout the interval. In the first case, the subjects have obviously detected the temporal nature of the contingency and probably resort to some mental chronometric device which occasionally helps time estimation; in the second case they misinterpret the situation as if it would require sustained responding. Such contrasting patterns appear to be cognitive interpretations, plausibly based on linguistic capacities. In the second group of subjects, the cognitive elaboration would result in a less adaptive performance than the one observed in animals. Comparison with animal subjects is not enough to decide whether language or some other human species-specific character is the crucial variable here. A developmental approach recommends itself to verify if the acquisition of language is responsible for the behavior observed in humans adults.

Developmental evidence

Developmental studies of psychological time have been mainly devoted, after the pioneering work of Piaget (1946), to the construction of the concept of time (Fraisse, 1948; 1982; Montangero, 1977) and to the progressive mastery of the linguistic tools expressing time relations (Bronckart, 1976; Ferreiro, 1971). By their very nature, such studies can only be carried out on subjects having already acquired language. The same applies to studies on time estimation proper, for reasons of experimental accessibility: the procedures used involve verbal instructions to the subjects. Operant procedures applied to infants or very young children have provided information which allow us to compare their temporal regulations to those of animals and to those of human adults, and consequently to better assess the role of language in the patterns characteristically observed in human adults.

In its turn, the comparative approach to infants and young children's performances requires reference to developmental data in animals. These have been, until recently, very scarce, not to say completely absent, reflecting a lack of interest for the heuristic and explanatory value of developmental research among experimenters in the operant tradition (for a discussion of that particular point in a wider context see Richelle, 1992; 1993). The few studies reported below are recent and therefore still exploratory. They will be briefly discussed before returning to infant studies and to the role of language.

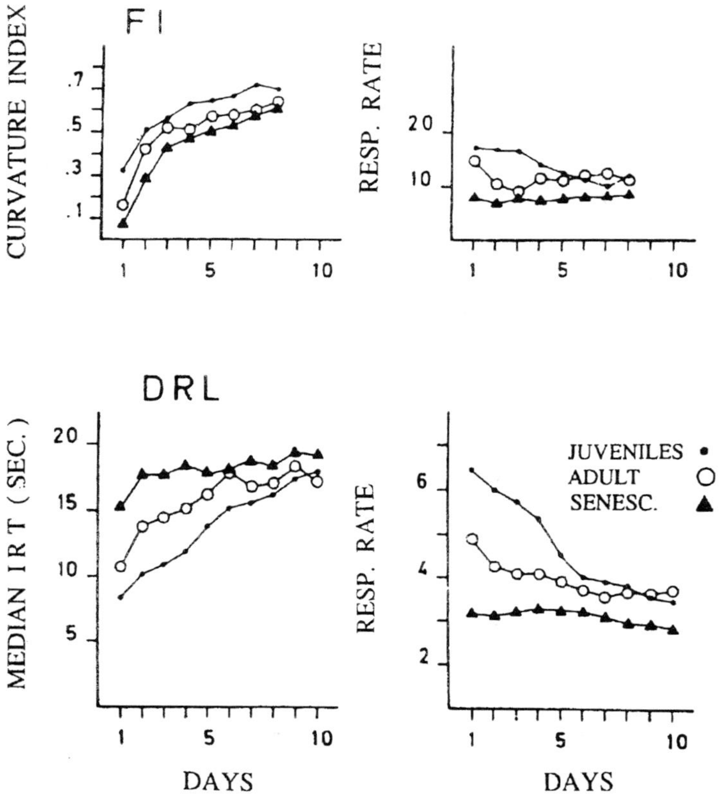

Figure 6: Top: curvature index (left) and response rate (right) in juvenile, adult and senescent rats during acquisition of behavior under a FI 60 seconds schedule over 10 days of training [the curvature index is an estimate of the quality of spontaneous temporal regulation in FI schedules]. Bottom: median IRT (left) and response rate (right) during acquisition of behavior under a DRL 20-seconds schedule over 10 days of training (after Lejeune, 1990)

Development studies in animals

With very few exceptions (such as Marley & Morse, 1966), animal developmental studies on temporal regulations do not go back more than 15 years. Some of them are based on a life-span developmental approach and as such they provide data on aging subjects as well. Methodological difficulties involved in longitudinal studies in very young and very old animals have been discussed in depth and appropriately solved. For example, to avoid food deprivation in young animals, and reduce the interaction between learning processes and maturational factors in fastly growing organisms, subjects were submitted to four or five daily sessions, instead of one, over a period of 8 to 10 days. Prematurely weaned rats (16 days) were conditioned, under those conditions, as early as 18 to 21 days of age (Lejeune, 1992). Also, nest-leaving birds, more amenable to early conditioning, such as quails or ducklings were used (Lejeune & Nagy, 1986; De Paulo & Hoffman, 1981).

Young (20 days old), adults (3 months old) and senescent (24 months old) rats were compared in FI 60-seconds and DRL 20-seconds schedules (Lejeune, 1989, 1992; Lejeune et al., 1986). Results are summarized in Figure 6 (p. 160). In the FI schedule, temporal regulation is best in young rats, in spite of a higher rate of responses, reflecting higher level of activity. In the DRL schedule, the response rate is similarly related to age, but in this case it has deteriorating effects on timing in young animals. Similar results were obtained with a schedule of differential reinforcement of response duration (DRR). In these cases, activity level interferes with timing. This could be nothing more than the aspecific effect of a trivial variable - „general activity" as a characteristic of age - comparable to motivation, emotional arousal, etc. However, this does not explain why and how spontaneous temporal regulations observed in FI do not suffer from general activity. This is a paradox that has been pointed to many years ago, and one that developmental studies show even more strikingly, and that until now no formal model of psychological time has explained in any satisfying way.

Whatever the contingencies used, differences between age groups fade out as training is extended in time, eventually to disappear after 40 to 50 sessions. This suggests that differences observed are not due to different processes of the internal clock - whatever it is. It also warns the experimenter working on human subjects not to indulge in hasty conclusions from results collected on children, human adults or older people in a few short experimental tests, as is generally the case.

Developmental studies in humans

Exploratory and incomplete as they are, these findings will frame the data collected in infants and children in both comparative and developmental perspective.

The performance of infants in FI does not resemble the adults' performance. Infants show the typical pattern observed in animals, that is a pause after reinforcement, the duration of which is proportional to the interval (Lowe et al., 1983). Between 2 1/2 years and 5 years, the distribution of responses throughout the interval is erratic (Droit et al., 1990). Typical patterns of human adults, described above, appear at around 5 years. This change could be due to the mastery of language, especially as a tool for self-description and self-instruction concerning the temporal contingencies and ways to cope with them (Bentall & Lowe, 1987; Bentall et al., 1985). Erratic patterns observed between 2 and 5 would essentially reflect the linguistic immaturity. These findings suggest a progressive shift from contingencies controlled to rule-governed behavior, the transition phase being marked by the competition of the two modes.

These interesting data, which fit nicely the course of cognitive and linguistic development, raise some problems of replication. Darcheville et al. (1993) have reported, in infants 3 to 5 months of age, in a FI schedules (using increasing intervals from 10 to 80 seconds) a very low response rate comparable to what is observed in those adults who seem to have detected the time contingencies. Babies age 9 to 13 months (comparable in age to the subjects of Lowe et al., 1983) have performances falling in one of the two categories described in adults. Individual patterns seem to correlate with the behavior observed in a self-control task, in which the subject has the choice between an immediate but short reinforcer or a differed but long one. Those babies which make the first choice exhibit high responding in the FI, while those which make the second choice show low rates. A number of methodological variables, such as the response used, the type of reinforcer, general aspects of the experimental situation, have to be checked before the debate can be closed. Nevertheless, the findings throw some doubt on the purely cognitive interpretation of infants' performances in terms of language acquisition, and may suggest that we more explore temperamental predisposition, a concept recently rehabilitated after a long period of disrepute (Kagan & Snidman, 1991). That line of reflection brings us back to the hypothesis of giving inhibitory processes a major place in temporal regulations.

Adaptation to DRL and DRRD schedules does not leave much freedom: only one pattern of behavior provides the reinforcer. This constraint is essential in understanding the development of performances in young children and young animals. Contrary to what is observed in FI schedules, no major difference appears in DRL or DRRD performances as a function of age. In both schedules, young children (and rats as well) characteristically produce a number of inadapted responses, which will progressively decrease as motor inhibitory functions develop (Lejeune & Jasselette, 1987a and b; Lejeune & Richelle, 1990). As it seems, the child becomes increasingly aware with age that it has to wait for a given duration (Pouthas, 1985; 1990; 1993). From its second year of age, it often fills the waiting time by producing collateral behavior (also observed in animals) of all sorts, sometimes requiring much motor energy, such as running around the room, turning its chair upside down, and the like. These collateral behaviors become less frequent with age (Pouthas, 1985; Pouthas & Jacquet, 1987). Some children at 4 years of age produce very accurate timing in DRL 5 s. Others take slightly longer, but eventually at the age of 7 performance is usually excellent.

It is obvious that human beings are confronted from birth with duration, through the frustrating experience of having to wait until their needs are satisfied. Waiting, which Janet (1928) rightly described as „active regulation of action," is the core of experimental procedures of time differentiation. They probably favor in *all* subjects the awareness of the role of time, while *some* subjects do eventually get aware of it in less constraining procedures, as FI schedules. The issue of awareness (or consciousness), which is again in favor in psychological and neuroscientists' circles, deserves systematic attention.

Conclusion

Attempts to develop general models of psychological time are the product of increasingly rich information collected during the last twenty years in various subfields of research devoted to various facets of time processing. They are based on the legitimate assumption that time processing can be described by a single model, however complex, and admittedly requiring qualifications to account for some of the peculiarities that might be observed at some levels of functioning. The scalar model has been taken as an example of such attempts, because it has been shown to encompass a number of data collected both in animals and humans, and it is open enough to accomodate adjustments, at least to some extent, as formal models evolve from information processing to connectionist networks and as empirical findings multiply.

The comparative and developmental studies discussed in the present paper offer a large quantity of such empirical data. Some of them seem to fit the model fairly well, confirming its validity, and corroborating its heuristic value. Others, however, are less easy to incorporate: either they simply do not fit the requirements of the model, or they just lay outside of its scope. In both cases, they raise a problem as to the generality of the model. In the first case, the solution can be looked for either in the adjustment of the model itself - which might become increasingly complex and qualified - or in the rejection of empirical evidence as irrelevant to a theory of „pure" psychological time, that is to say, data not fitting the model would be explained away as effects of non-temporal variables. In the second case, empirical findings raise different questions, which the model does not address, but which, nevertheless, may be sound questions in the frame of a theory of time. They will have to be accounted for some day. Perhaps a new, enlarged, model will be needed. Maybe new models will do better, accounting for the real diversity of processes involved in psychological time, which seem to challenge the search for a single unified theory that seems so attractive to current students of time.

References

Allan, L. (1992). The internal clock revisited. In F. Macar, V. Pouthas, & W.J. Friedman (Eds.), *Time, action and cognition: towards bridging the gap* (pp. 191-202). Dordrecht: Kluwer Academic Publishers.

Bentall, R.P., & Lowe, C.F. (1987). The role of verbal behavior in human learning: III Instructional effects in children. *Journal of the Experimental Analysis of Behavior, 47,* 177-190.

Bentall, R.P., Lowe, C.F., & Beasty, A. (1985). The role of verbal behavior in human learning: II Developmental differences. *Journal of the Experimental Analysis of Behavior, 43,* 165-181.

Bronckart, J.P. (1976). *Genèse et organisation des formes verbales chez l'enfant.* Bruxelles: Dessart et Mardaga.

Catania, A.C. (1970). Reinforcement schedules and psychophysical judgment: A study of some temporal properties of behavior. In W. N. Schoenfeld (Ed.), *Theory of reinforcement schedules* (pp. 1-42). New York: Appleton Century Crofts.

Cheng, K., Westwood, R., Crystal, J.D. (1993). Memory variance in the peak procedure of timing in pigeons. *Journal of Experimental Psychology: Animal Behavior Processes, 19,* 68-76.

Church, R.M. (1984). Properties of the internal clock. In J. Gibbon & L. Allan (Eds.), *Timing and time perception* (pp. 566-582). New York: Annals of the New York Academy of Sciences, 423.

Church, R.M. (1989). Theories of timing behavior. In S. B. Klein & R. R. Mowrer (Eds.), *Contemporary learning theories: instrumental conditioning theory and the impact of biological constraints on learning* (pp. 41-71). Hillsdale, NJ: Erlbaum.

Church, R.M., & Broadbent, H.A. (1990). Alternative representations of time, number and rate. *Cognition, 37*, 55-81.

Church, R.M., Meck, W.H. & Gibbon, J. (1994). Application of scalar timing theory to individual trials. *Journal of Experimental Psychology: Animal Behavior Processes, 20*, 135-155.

Darcheville, J.C., Rivière, V., & Wearden, J.H. (1993). Fixed interval performance and self-control in infants. *Journal of the Experimental Analysis of Behavior, 60*, 239-254.

De Paulo, P. & Hoffman, H.S. (1981). Reinforcement by an imprinting stimulus versus water on simple schedules in ducklings. *Journal of the Experimental Analysis of Behavior, 36*, 151-169.

Droit, S., Pouthas, V., & Jacquet, A.Y. (1990). Temporal learning in 4 1/2- and 6-year-old children: role of instructions and prior knowledge. *Journal of Experimental Child Psychology, 50*, 305-321.

Ferreiro, E. (1971). *Les relations temporelles dans le langage de l'enfant*. Genève: Droz.

Fraisse, P. (1948). Etude comparée de la perception et de l'estimation du temps chez les enfants et les adultes. *Enfance, 1*, 199-211.

Fraisse, P. (1982). The adaptation of the child to time. In W.J. Friedman (Ed.), *The developmental psychology of time* (pp. 113-139). New York: Academic Press.

Gibbon, J. (1977). Scalar expectancy theory and Weber's law in animal timing. *Psychological Review, 84*, 279-325.

Gibbon, J., & Church, R.M. (1990). Representation of time. *Cognition, 37*, 23-54.

Gibbon, J. & Church, R.M. (1992). Comparison of variance and covariance patterns in parallel and serial theories of timing. *Journal of the Experimental Analysis of Behavior, 57*, 393-406.

Janet, P. (1928). *L'évolution de la mémoire et de la notion de temps*. Paris: A. Chahine.

Jasselette, P., Lejeune, H., & Wearden, J.H. (1990). The perching response and the laws of animal timing. *Journal of Experimental Psychology: Animal Behavior Processes, 16*, 150-161.

Kagan, J. & Snidman, N. (1991). Temperamental factors in human development. *American Psychologists, 46*, 856-862.

Leak, T.M. & Gibbon, J. (1995). Simultaneous timing of multiple intervals: Implications of the scalar property. *Journal of Experimental Psychology: Animal Behavior Processes, 21*, 3-19.

Lejeune, H. (1971). Note sur les régulations temporelles acquises en programme à intervalle fixe chez le chat. *Revue du Comportement Animal, 5*, 123-129.

Lejeune, H. (1989). Long term memory for DRL: a comparison between weanling, adult and senescent rats. *Physiology and Behavior, 45*, 321-330.

Lejeune, H. (1990). Timing: generality beyond differences or differences in continuity? In D. Blackman & H. Lejeune (Eds.), *Behaviour analysis in theory and practice: contributions and controversies* (pp. 53-90). London: Erlbaum.

Lejeune, H. (1992). Response timing and development : fixed-interval performance in precociously weaned rats. *The Quarterly Journal of Experimental Psychology : Comparative and Physiological Psychology, 44B*, 101-122.

Lejeune, H. (1993). *Estimation des durées et régulation des conduites: faits et modèles*. Université de Liège: Faculté de Psychologie, P.A.I.

Lejeune, H., & Jasselette, P. (1987a). DRL performance in the weanling rat: A comparison with adult subjects. *Physiology and Behavior, 40*, 271-278.

Lejeune, H., & Jasselette, P. (1987b). Differential reinforcement of response duration (DRRD) in weanling rats: a comparison with adult subjects. *Behavioural Processes, 15*, 315-332.

Lejeune, H., Jasselette, P., Nagy, J. & Perée, F. (1986). Fixed Interval performance in the weanling rat: a comparison with adult and senile subjects. *Physiology and Behavior, 38*, 337-343.

Lejeune, H., & Macar, F. (1994). Régulations temporelles. In M. Richelle, J. Requin, & M. Robert (Eds), *Traité de Psychologie Expérimentale*. Paris: Presses Universitaires de France.

Lejeune, H., & Nagy, J. (1986). Operant conditioning in the newly hatched quail: fixed interval performance. *Behavioural Processes, 12*, 317-326.

Lejeune, H., & Richelle, M. (1982). Differential reinforcement of perching duration in the pigeon: A comparison with differential reinforcement of low rate key pecking. *Behaviour Analysis Letters, 2*, 49-57.

Lejeune, H., & Richelle, M. (1990). Timing behavior and development: comments on some animal and human data. *International Journal of Comparative Psychology, 4*, 111-135.

Lejeune, H. & Soffié, M. (in preparation). Peak procedure in the aged rat: behavioral flexibility to changes of the temporal criterion.

Lejeune, H., & Wearden, J.H. (1991). The comparative psychology of fixed interval performance: some quantitative analysis. *Learning and Motivation, 22*, 84-111.

Lowe, C. F., Beasty, A., & Bentall, R. P. (1983). The role of verbal behavior in human learning: infant performance on fixed interval schedules. *Journal of the Experimental Analysis of Behavior, 39*, 157-164.

Macar, F., Grondin, S. & Casini, L. (1994). Controlled attention sharing influences time estimation. *Memory & Cognition, 22*, 673-686.

Mantanus, H. (1981). Empty and filled interval discrimination by pigeons. *Behaviour Analysis Letters, 1*, 217-224.

Marley, E., & Morse, W.H. (1966). Operant conditioning in the newly hatched chicken. *Journal of the Experimental Analysis of Behavior, 9*, 95-103.

Mechner, F. & Guevrekian, L. (1962). Effects of deprivation upon counting and timing in the rat. *Journal of the Experimental Analysis of Behavior, 6*, 463-466.

Meck, W.H. (1984). Attentional bias between modalities: Effects on the internal clock, memory, and decision stages used in animal time discrimination. In J. Gibbon, & L. Allan (Eds.), *Timing and time perception* (pp. 528-541). New York: Annals of the New York Academy of Sciences, 423.

Meck, W.H., & Church, R.M. (1982). Abstraction of temporal attributes. *Journal of Experimental Psychology: Animal Behavior Processes, 8*, 226-243.

Meck, W.H., Church, R.M. & Wenck, G.L. (1986). Arginine vasopressin inoculates again age-related increases in sodium-dependent high affinity choline uptake and discrepancies in the content of temporal memory. *European Journal of Pharmacology, 130,* 327-331.

Michon, J. (1990). Implicit and explicit representations of time. In R.A. Block (Ed.), *Cognitive models of psychological time* (pp. 37-58). Hillsdale, NJ: Erlbaum.

Montangero, J. (1977). *La notion de durée chez l'enfant de 5 à 9 ans*. Paris: Presses Universitaires de France.

Piaget, J. (1946). *Le développement de la notion de temps chez l'enfant.* Paris: Presses Universitaires de France.

Platt, J. R. (1979). Temporal differentiation and the psychophysics of time. In M.D. Zeiler, & P. Harzem (Eds.), *Advances in the analysis of behaviour, Vol. 1: reinforcement and the analysis of behaviour* (pp. 1-29). Chichester: Wiley.

Pouthas, V. (1985). Timing behavior in young children: a developmental approach to conditioned spaced responding. In J. Michon, & J. Jackson (Eds.), *Time, mind and behavior* (pp. 100-109). Berlin: Springer.

Pouthas, V. (1990). Temporal regulation of behaviour in humans: A developmental approach. In D. Blackman, & H. Lejeune (Eds.), *Behaviour analysis in theory and practice: contributions and controversies* (pp. 33-51). London: Erlbaum.

Pouthas, V. (1993). Ontogenesis of temporal learning in the child: experimental evidence and perspectives. *Psychologica Belgica, 33,* 171-184.

Pouthas, V., & Jacquet, A. Y. (1987). A developmental study of timing behavior in 4 1/2- and 7-year-old children. *Journal of Experimental Child Psychology, 43*, 282-299.

Richelle, M. (1968). Notions modernes de rythmes biologiques et régulations temporelles acquises. In J. de Ajuriaguerra (Ed.), *Cycles biologiques et psychiatrie*. (pp. 233-255). Paris: Masson.

Richelle, M. (1972). Temporal regulation of behaviour and inhibition. In R.A. Boakes & M. S. Halliday (Eds.), *Inhibition and learning* (pp. 229-251). London: Academic Press.

Richelle, M. (1992). From action to cognition: Bridging the gap? In F. Macar, V. Pouthas, & W.J. Friedman (Eds.), *Time, action and cognition: towards bridging the gap* (pp. 373-382). Dordrecht: Kluwer Academic Publishers.

Richelle, M. (1993). A concise guide to the temporal maze. *Psychologica Belgica* (Special issue on „Temporal processes“), *33*, 135-145.

Richelle, M. & Lejeune, H. (Eds.) (1980). *Time in animal behaviour*. Pergamon Press.

Richelle, M., & Lejeune, H. (1984). Timing competence and timing performance: A comparative approach. In J. Gibbon, & L.G. Allan (Eds.), *Timing and time perception* (pp. 254-268). New York: Annals of the new York Academy of Sciences, 423.

Richelle, M., Lejeune, H., Fery, P., & Perikel, J.J. (1985). From biotemporality to nootemporality: toward an integrative and comparative view of time in behavior. In J.A. Michon & J.L. Jackson (Eds.), *Time, mind, and behavior*. (pp. 75-99). Berlin: Springer.

Roberts, S. (1981). Isolation of an internal clock. *Journal of Experimental Psychology: Animal Behavior Processes, 7*, 242-268.

Roberts, S., Cheng, K., & Cohen, J. S. (1989). Timing light and tone signals in pigeons. *Journal of Experimental Psychology: Animal Behavior Processes, 15*, 23-35.

Roberts, S. & Holder, M.D. (1984). What starts an internal clock? *Journal of Experimental Psychology:, 10,* 273-296.

Stubbs, D.A. (1979). Temporal discrimination and psychophysics. In M.D. Zeiler, & P. Harzem (Eds.), *Advances in the analysis of behaviour, Vol. 1: Reinforcement and the organization of behaviour* (pp. 341-369). Chichester: Wiley.

Treisman, M. (1963). Temporal discrimination and the indifference interval: Implications for a model of the „internal clock.“ *Psychological Monographs, whole n° 576*.

Wasserman, E.A. (1993). Comparative cognition: beginning the second century of the study of animal intelligence. *Psychological Bulletin, 113*, 211-228.

Wearden, J.H. (1985). The power law and Weber's law in fixed-interval postreinforcement pausing: A scalar timing model. *Quarterly Journal of Experimental Psychology: Comparative and Physiological Psychology, 37B*, 191-211.

Wearden, J.H. (1991a). Do humans possess an internal clock with scalar timing properties ? *Learning and Motivation, 22*, 59-83.

Wearden J.H. (1991b). Maximizing reinforcement rate on spaced responding schedules under conditions of temporal uncertainty. *Behavioural Processes, 22*, 47-59.

Wearden, J.H., & Lejeune, H. (1993). Across the great divide: animal psychology and time in humans. *Time and Society, 2*, 87-106.

Wilkie, D.M. (1987). Stimulus intensity affects pigeons' timing behavior: implications for an internal clock model. *Animal Learning and Behavior, 15*, 35-39.

Zakay, D. (1989). Subjective and attentional resource allocation: an integrated model of time estimation. In I. Levin & D. Zakay (Eds.), *Time and human cognition* (pp. 365-397). North-Holland: Elsevier.

Zeiler, M.D. (1983). Integration of response timing: the functional time requirement. *Animal Learning and Behavior, 11*, 237-246.

Zeiler, M.D. (1985). Pure timing in temporal differentiation. *Journal of the Experimental Analysis of Behavior, 43*, 183-193.

Zeiler, M.D. (1986). Behavior units and optimality. In T. Thompson & M.D. Zeiler (Eds.), *Analysis and integration of behavior units* (pp. 81-116). Hillsdale; N.J.: Erlbaum.

Chapter 9: Models of psychological time revisited*

RICHARD A. BLOCK and DAN ZAKAY

Introduction

Theorists have taken two seemingly different approaches to explaining, or modeling, psychological time (Block, 1990). These approaches have appeared under several guises. Ornstein (1969) referred to them as the sensory-process approach and the cognitive approach. Sensory-process models „postulate some sort of 'time-base', a repetitive, cumulative, pulse-dispensing mechanism which delivers internal time signals, an 'organ' of time" (p. 25). Ornstein claimed that this type of model has not provided a useful way to understand duration experience. This approach may also have difficulties explaining why cognitive, or information-processing, variables influence duration experience. The other class of model includes various proposals concerning the important cognitive factor underlying duration experience, such as „images" (Guyau, 1890; 1988), „changes," (Fraisse, 1957; 1963), „mental content" (Frankenhaeuser, 1959), „storage size" (Ornstein, 1969), and „contextual changes" (Block & Reed, 1978). According to some proponents of sensory-process models, cognitive models cannot easily explain the near-linear psychophysical relationship between physical and psychological duration, as well as the possible influence of physiological variables such as body temperature.

Timing with or without a timer

The important difference between the two approaches is not that the first concerns sensory processes and that the second concerns cognitive processes. Instead, the first class of model proposes timing with a timer, whereas the second proposes timing without a timer (Ivry & Hazeltine, 1992). In timing-with-a-timer models, a pacemaker mechanism underlies

* We thank Hannes Eisler, Françoise Macar, John Moore, and Andras Semjen for helpful comments on an earlier draft of this chapter.

the psychological timing system. Two major variants are chronobiological and internal-clock models (Block, 1990).[1] In timing-without-a-timer models, subjects construct psychological time from processed and stored information - that is, some salient aspect or byproduct of information processing. Variants of this kind of model include attentional, memory storage, and memory change models (Block, 1990).

Piéron (1923) was one of the first researchers to discuss the possible relationship between body temperature and duration experience. Subsequently, François (1927) and Hoagland (1933) obtained such evidence, which supports a possible timing-with-a-timer model. Hoagland (1933; see also 1966) proposed that a *master chemical clock*, or temporal pacemaker, in the brain regulates time-related behaviors and judgments. The evidence suggested that the rate of repeated time productions - involving counting at the rate of one per second - increases as a function of body temperature. More recent evidence suggests that duration judgments of many minutes (e.g., hourly productions) are correlated with body temperature (Campbell & Birnbaum, 1994). Although the relationship between body temperature and shorter duration judgments is often inconsistent (Hancock, 1993), changes in body temperature do seem to lead to systematic changes in the rate of psychological time (Wearden & Penton-Voak, 1995). One possibility is that body temperature influences general arousal level, which thereby influences the rate of a pacemaker mechanism (Wearden & Penton-Voak, 1995). The problem with postulating that a pacemaker or master biochemical clock directly influences time-related behaviors and judgments is that temperature may also influence brain processes that subserve attentional, memory, and other cognitive processes. Variations in these processes probably have little or no effect on body temperature. Because cognitive variables (e.g., attentional demands of a task) influence duration experience, cognitive processes may directly mediate temporal behaviors and judgments. Body temperature may indirectly influence temporal behaviors and judgments by altering whatever cognitive processes subserve psychological time (Block, 1990).

Theorists have proposed a large number of cognitive models of psychological time. They have stated these mostly in the form of a „variable-x hypothesis," where one may substitute any of several variables for *variable-x* (e.g., input segmentation, complexity-of-coding, attentional selectivity). Each of these variables is typically the only one that the researcher manipulated. A few models have attempted to be

[1] An example of an internal-clock model is also desribed in chapter 4 by H. Eisler.

more general. Consider, for example, Doob's (1971) model (Figure 1, p. 174). This model illustrates a taxonomy of time that depicts interactions involving the „intricate, multivariate phenomenon of time" (p. 30). The details are relatively unimportant. For present purposes, we note that although this model may seem comprehensive, it is not a functional (e.g., information-processing) model of temporal behavior or judgment.

Block (1985) proposed a contextualistic model in which temporal experience is a product of four kinds of interacting factors (Figure 2, p. 175). Again, the details are relatively unimportant, because this model is only a little more functionalistic than Doob's. The main advantage of models such as Doob's (1971) and Block's (1985) is heuristic: these models remind us that psychological time involves complex interactions of various organismic and environmental variables. The main disadvantage of these models, as noted above, is that they do not „relate in a functional way to the empirical findings [they are] supposed to represent" (Michon, 1985, p. 26). Although both models depict interactions of variables, several functional issues remain: (a) Which interactions are important in particular situations and which are not? (b) What is the nature of the higher-order interactions? (c) How are the underlying processes sequenced, as in a functionalistically oriented information-processing model of temporal behavior?

Cognitive psychologists and others have occasionally proposed models resembling internal-clock models, but these usually involve timing without a timer. For example, Lashley (1951) thought that practiced movement sequences are structured as individual elements organized into chunks which are executed as part of a motor program for the action sequence. Because he proposed that a motor program is executed without the need for feedback, it needs an internal-control process to time elements. Researchers have searched for such a common mechanism that is able to stabilize motor programs despite changes in states of the organism, changes in contextual stimuli, changes in equipment or instruments used for the performance, and so on. The important question of how movement sequences are timed is still largely unresolved, as is the question of whether we need to propose an internal-clock mechanism. Motor programs may contain internal, hierarchically organized information about timing relationships, so a pacemaker mechanism may be unnecessary. Alternatively, even such information about timing relationships may rely on a pacemaker for some basic calibration (see Semjen, this volume, chapter 2).

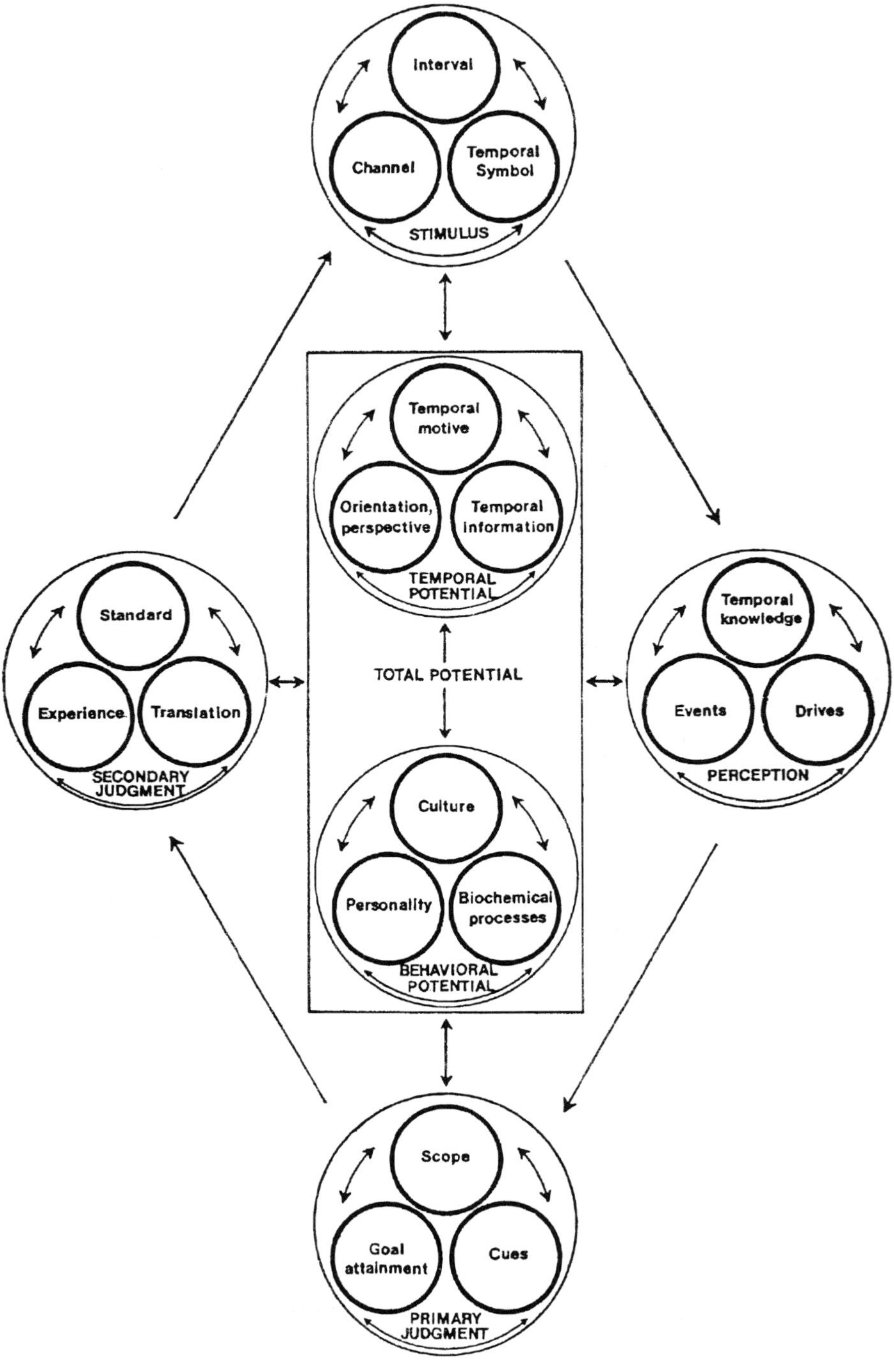

Figure 1: Doob's (1971) taxonomy of time.

In the remainder of the chapter, we review various formal models of psychological time. We propose the attentional-gate model, which reconciles the two approaches. This model is somewhat isomorphic with contextual-change models of experienced and remembered duration.

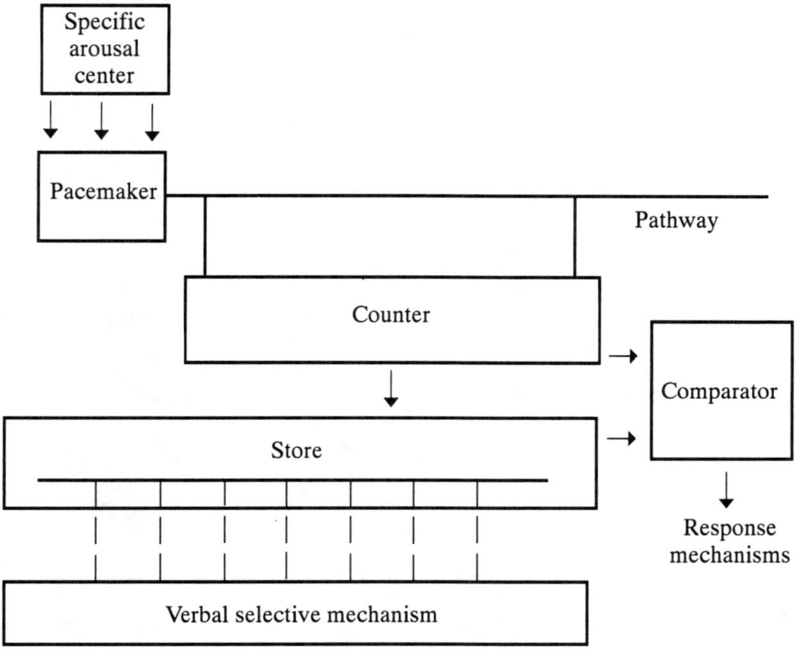

Figure 2: Block's (1985) contextualistic model of duration experience.

Treisman's Model

Treisman (1963) proposed an influential model of an internal clock underlying human temporal judgment (Figure 3, p. 176). He postulated a *pacemaker* that produces a regular series of pulses, the rate of which varies as a function of input from an organism's *specific arousal center*. In his view, specific arousal is influenced by external events, in contrast to general arousal, which depends on internal mechanisms such as those underlying circadian rhythms. A *counter* records the number of pulses in a *pathway*, and the total is transferred into a store and into a *comparator* mechanism. A *verbal selective mechanism* assists in retrieving useful information from the store. This is presumably a long-term memory store containing knowledge of correspondences between total pulses and verbal labels, such as 20 s, 1 m, and so on.

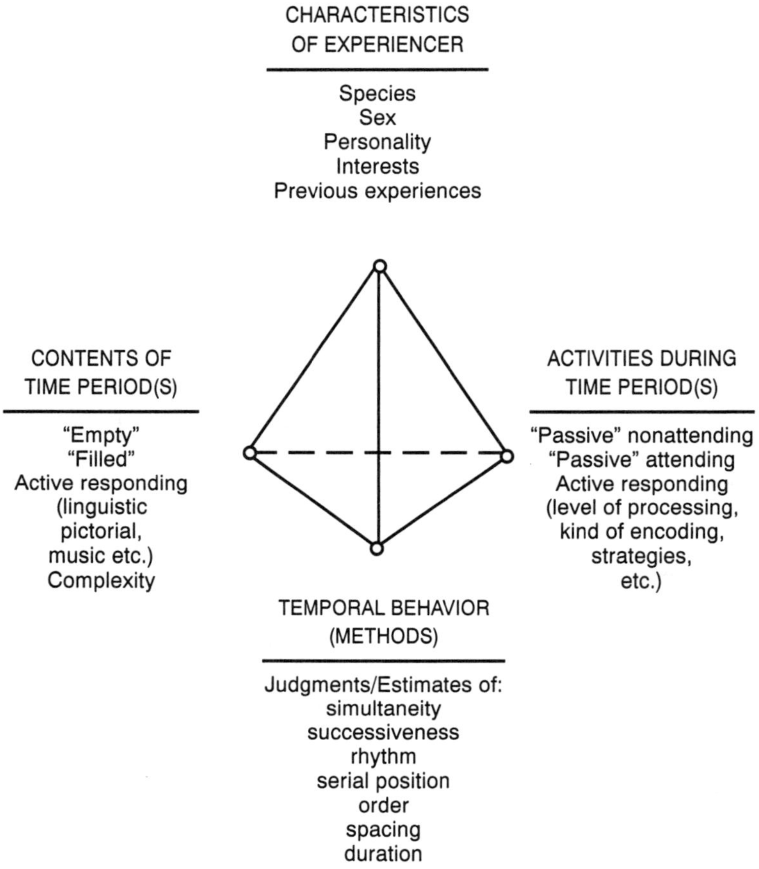

Figure 3: Treisman's (1963) model of the internal clock.

Subsequently, several theorists proposed that the alpha rhythm may reflect the frequency of the pacemaker component of the hypothetical internal clock. Treisman (1984) attempted to determine whether this is the case. However, his data do not support the view that arousal, at least as it is reflected in alpha frequency, influences the pacemaker rate. In a recent modification of this internal-clock model, Treisman and his colleagues (Treisman et al., 1990; Treisman, 1993) proposed a more complex pacemaker, which includes a *calibration unit* that can modulate the pulse rate. In our view, this minor modification does not rectify the limitations inherent in the model.

Scalar-timing model

Contemporary behavioral psychologists, who draw and test inferences about timing processes in animals, have proposed internal-clock models that resemble Treisman's (1963; 1993) model. These researchers typically investigate time-related behavior of animals such as pigeons and rats during relatively short time periods (seconds to minutes). The general finding is that animals are sensitive to different stimulus durations and time-based schedules of reinforcement.

Figure 4 (p. 177) shows the canonical model embodying the theory underlying these explanations, called *scalar timing theory* or *scalar expectancy theory*. Because this model provides an excellent account of a wide variety of evidence, many researchers have adopted it (e.g., Allan, 1992; Church, 1984; Gibbon & Church, 1984; Gibbon et al., 1984; Roberts, 1983). The present account is rather brief. We recommend Church's (1989) excellent chapter for additional details, and also Lejeune & Richelle's chapter (this volume, chapter 8), which contains an especially valuable discussion of cross-species comparisons.

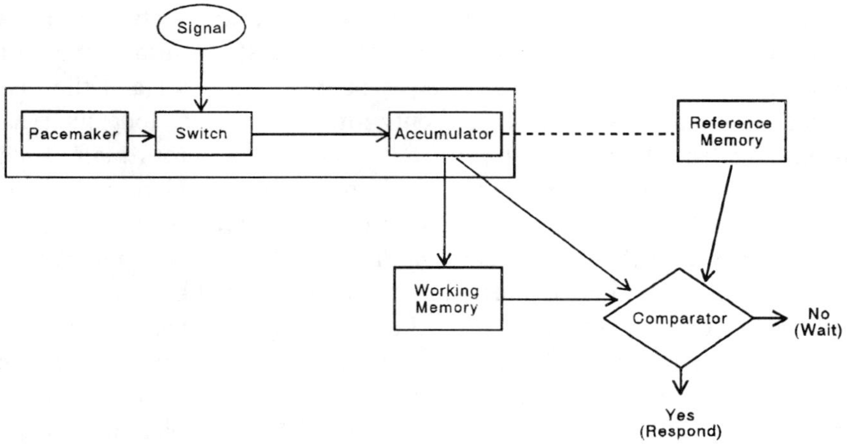

Figure 4: Scalar timing model (Church, 1984; Gibbon, 1984).

This model accounts for time (duration) perception and time production by proposing an internal clock, memory stores, and a decision mechanism. The internal-clock consists of a pacemaker, a switch, and an accumulator. The *pacemaker*, operating like a metronome, automatically and autonomously generates more or less regularly spaced pulses (at a rate of Λ pulses per second). When the organism perceives an external timing *signal* indicating the beginning of a time period, the *switch* operates (perhaps with a slight lag, influenced by attention), thereby allowing pulses to pass through to an accumulator. The *accumulator* integrates and holds the total pulse count during the time period (Λt). Perceived duration is a monotonic function of the total number of pulses transferred into the accumulator. On any trial, the contents of the accumulator are transferred into a *working memory* store for comparison with the contents of the *reference memory* store. The *reference memory* store contains a long-term memory representation of the approximate number of pulses that accumulated on past trials. This number is then transferred to the comparator with some bias, K^*, a memory storage constant that may be slightly less than or greater than 1. The *comparator* compares the contents (total pulse count) of the two stores.

Animal evidence

The peak procedure, which uses a modified discrete-trials fixed-interval (FI) schedule, is a common method used to explore animal timing. A relatively long, variable interval separates each trial. The onset of a discriminative stimulus such as a light signals the start of each trial. On most trials, the first response occurring after a FI (e.g., 30 s) has elapsed since the start of the trial is reinforced; then the discriminative stimulus is turned off. On other trials, which are the most important ones for testing the theory, the animal receives no reinforcement, and the discriminative stimulus is turned off only after a relatively long interval, usually at least twice the FI (e.g., 60 s). Averaged across many such trials, the typical response rate is approximately a Gaussian (bell-shaped) function of time since the start of the interval. This timing behavior reveals a scalar property: regardless of FI length, the average response rate at any time, expressed as a proportion of the peak rate, is a function of the proportion of the total duration that has elapsed. In other words, the normalized response-rate curve does not vary much from one FI length to another. The internal clock model handles this general finding by proposing that the response rate increases in probability as the comparison of working memory and reference memory reveals a similar total pulse count.

In the peak procedure and other similar procedures, the switch operates at the onset of the discriminative stimulus, with some slight lag attributable to attentional processes, and the accumulation of pulses begins. If an animal learns that the temporary offset of the timing signal will delay reinforcement by the length of the offset duration, it shows appropriate responding - that is, a temporally displaced response rate function (Roberts & Church, 1978). This implies that the internal clock functions like a stopwatch, cumulatively timing the duration. The start/stop button may be switched off for the duration of the offset of the timing signal. (This represents a slight elaboration on Treisman's model, which did not propose a counter-stopping mechanism.)

Human evidence

Because the scalar-timing model can time various periods, including interrupted fixed-interval schedules, it is able to handle virtually all extant animal-timing data. However, the model does not take into account factors that are more prominent in humans than in other animals. In particular, it is not easily able to explain why cognitive factors (e.g., attention, strategies, information-processing tasks) influence temporal behaviors. This seems largely a consequence of methodological limitations or neglect: few animal timing researchers have explored or discussed the effects of attentional manipulations, which have been a focus of considerable research on human prospective duration timing. In addition, organisms may use repetitive or chained behaviors as „external clocks" to time intervals; that is, they may engage in movements for an appropriate amount of time while they wait for reinforcement to be enabled (Pouthas, 1985). Thus, activities (such as strategies) of an organism during a time period influence its time-related behaviors. The scalar-timing model does not incorporate this kind of „external" timing process.

In short, internal-clock models proposed by behavioral psychologists investigating timing in nonhuman animals seem somewhat limited (Block, 1990). Until these models consider the role of cognitive factors, such as attentional allocation, they will not be able to generalize to explaining human duration judgment. The work of Richelle and Lejeune (see Lejeune & Richelle, this volume, chapter 8; Richelle & Lejeune, 1980; Richelle et al., 1985) is a notable exception. They have conducted comparative research involving several species, including humans, and have included a role for cognitive factors. Richelle et al. (1985) thought that the answer to the problem is to propose „as many clocks as there are behaviors exhibiting timing properties" (p. 90). We take a different view. We propose a model containing a single „cognitive clock," a model with seemingly broad explanatory power (see „attentional-gate model," later in this chapter).

To justify such a model, we first consider the more cognitively oriented human research. We show that without additional elaboration the scalar-timing model cannot handle evidence on human timing and temporal judgment.

Thomas's attentional model

Several theorists have proposed models of psychological time in which attention to time, or temporal information processing, plays a major role. Thomas and his colleagues (Thomas & Cantor, 1975; Thomas & Weaver, 1975) developed and tested a mathematical model in which attentional allocation influences duration judgments. This model (Figure 5, p. 181) is the most explicitly formulated attentional model of psychological time. It is expressed as the functional equation: $\tau(I) = \alpha f(t, I) + (1 - \alpha) g^*(I)$. The model says that the perceived duration τ of an interval containing certain information I is a monotonic function of the weighted average of the amount of information encoded by two processors, a temporal information processor f(t, I), and a nontemporal information processor g*(I). The organism divides attention between the two processors, which function in parallel. Perceived duration is weighted (with probability parameter α) to optimize the reliability of the information that each processor encodes, because as more attention is allocated to one processor, the other becomes more unreliable. As α approaches 1, the subject encodes more temporal information, and as α approaches 0, the subject encodes more nontemporal information. If less stimulus information occurs during the to-be-judged duration, the organism allocates more attention to temporal information, and f(t, I) is more heavily weighted. If a task demands more information processing, the organism allocates more attention to this nontemporal information, and g*(I) is more heavily weighted.

Although Thomas and his colleagues studied only human duration judgments of stimuli presented for less than 100 ms, one may consider the model to be a general model of temporal information processing, even involving longer time periods (Michon, 1985).

The strengths of this model complement the strengths of the scalar-timing model. Extant formulations and empirical tests of the scalar-timing model only use the concept of attention in a very limited way (see, for example, Allan, 1992; Meck, 1984). Because attention plays a critical role in Thomas's model, it handles an aspect of the human data that scalar-timing models do not. It also makes slightly more precise terminology such as *attention to time* and *temporal information processing*, which Block (1990) criticized as being overly vague. The model adds precision to these terms by implying that attending to time is like attending to stimulus information in that both processes require access to the same limited resources. However, Thomas's model is deficient in handling the animal-timing data, as well as the role of physiological (noncognitive) factors. Because it assumes a constant pool of attentional resources, it does not consider arousal level or variations in alertness attributable to circadian rhythms and other biological factors. It is also too passive: Thomas proposed that stimulus information alone determines the allocation of attention and that strategies are not involved. The model needs a concept of attention along the lines of Kahneman's (1973) resource model. Kahneman argued that arousal determines the total attentional resources available at any moment to meet information-processing demands. Thus, temporal information processing is influenced not only by characteristics of the information-processing task, but also by momentary arousal level and, hence, total available resources. We need this modification of Thomas's model to handle new findings such as the fact that increased alertness, such as when a person is under the influence of stimulants like methamphetamine, lengthens duration experience (Frankenhaeuser, 1959; Hicks, 1992).

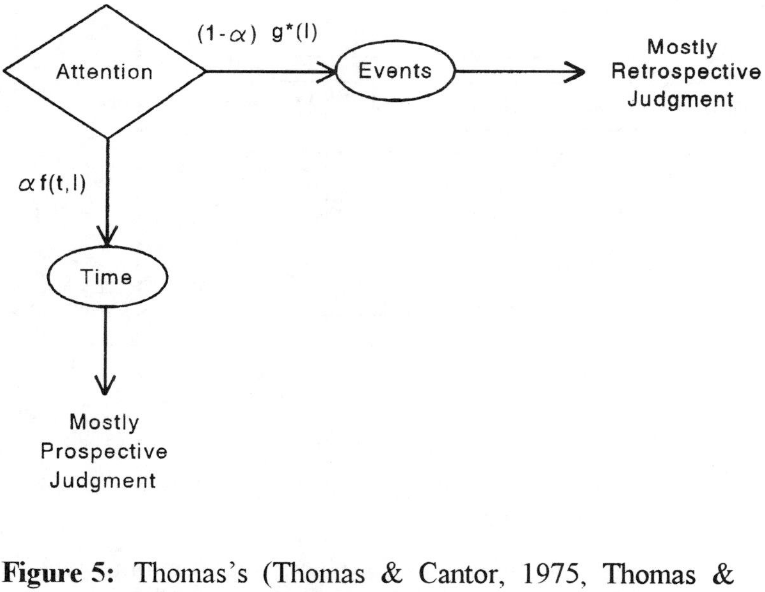

Figure 5: Thomas's (Thomas & Cantor, 1975, Thomas & Weaver, 1975) functional equation diagrammed as a model.

Attentional-gate model

We propose a model combining features of Treisman's (1963) model, the scalar-timing model (Church, 1984; Gibbon & Church, 1984; Gibbon et al., 1984), and Thomas's (Thomas & Cantor, 1975, Thomas & Weaver, 1975) model. We call this the *attentional-gate model* (Figure 6, p. 182). Consider first a version of the model that can handle the kind of prospective duration timing which both Treisman's and scalar-timing models are designed to handle. The critical feature of prospective timing is that the organism's behavior is focused on temporal information, as a result of either learning or instructions.

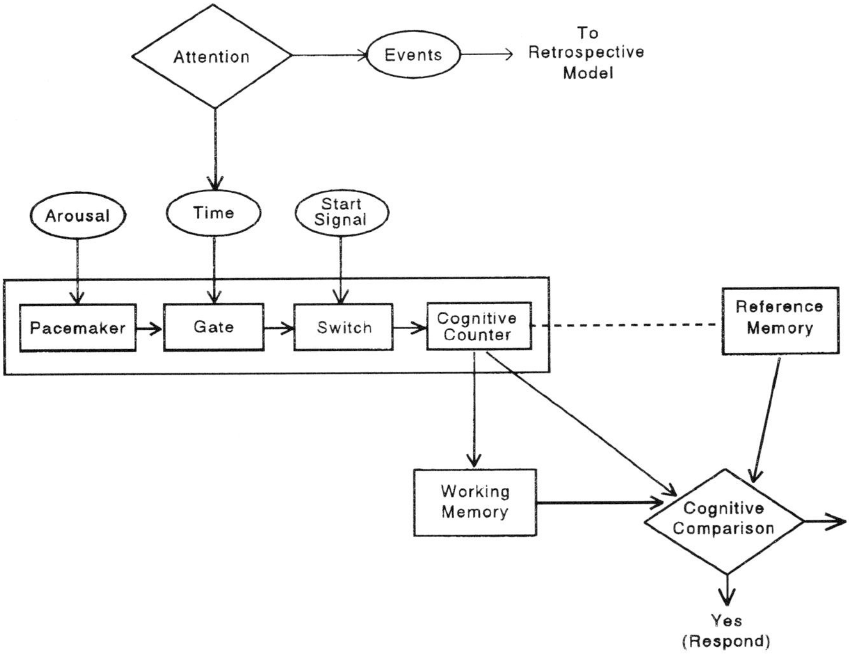

Figure 6: The attentional-gate model of prospective duration timing.

We propose that a *pacemaker* produces pulses at a rate which is influenced by both general (e.g., circadian) and specific (e.g., stimulus-induced) *arousal*. Each occasion on which an organism attends to time, as opposed to external stimulus events, opens a *gate*. This allows the pulse stream to be transmitted to subsequent components. At the onset of a duration, indicated by some *start signal*, the switch allows the pulse stream to be transmitted to the *cognitive counter*, where pulses are counted, or summed over time. (We call this component a *cognitive counter*, rather than simply an *accumulator*, because controlled cognitive processes, such as attention, influence the input to it.) The rest of the model contains functional components analogous to those in the scalar-timing model. The momentary total pulse count in the cognitive counter is transferred to a *working memory* store. (This process may occur only when attention is deployed, in contrast to the analogous process in the scalar-timing model, which is assumed to be automatic and continuous.) In addition, a *reference memory* store contains a record of the average total number of pulses that accumulated in the past before a certain time period was complete. (In humans, the reference memory store may also contain learned correspondences between pulse totals and verbal labels for temporal units.) If the momentary total pulse count in working memory approximates the total in reference memory, a *cognitive comparison* (probably also involving attention) results in the organism signaling the end of the time period or making some other duration-dependent response. If fewer than the required number of total pulses have been counted, the organism waits or makes a shorter duration judgment.

At present, little or no evidence (whether from animals or from humans) definitively tests some details of this model. For this reason, we are unsure about the relative location of two components, the attentional gate and the switch (Zakay & Block, 1994). It may be more appropriate to locate the switch before, instead of after, the attentional gate. Neither logical analysis nor empirical evidence seems to favor one order over the other. Differences in the dynamics of these two components suggests that they are separate components, instead of simply being a single *attentional switch* (cf. Allan, 1992; Meck, 1984): The switch operates as a result of the organism's processing of external signals, whereas the gate operates as a result of the organism's internal allocation of attentional resources. We are also unsure about the appropriate metaphor to use for the functioning of the gate. Attention to time may be viewed as opening the gate wider or more frequently, thereby allowing more pulses to pass through it to the cognitive counter. Neither logic nor evidence is available to distinguish these metaphors.

The attentional-gate model contains two important modifications to extant internal-clock models. First, it incorporates the notion that a subject may divide attentional resources between attending to external events and attending to time (Thomas & Cantor, 1975; Thomas & Weaver, 1975), and it specifies the consequences of each. Attending to time is necessary for pulses to be transmitted to the cognitive counter. While the duration is in progress (i.e., while the switch is allowing pulses to pass to the cognitive counter), the number of transmitted pulses is a function of two factors: (a) the pulse rate, which is influenced by general and specific arousal, and (b) the proportion of time the gate is open, or the width it is open, which is influenced by the amount of attention allocated to time.

Few animal-timing experiments have presented varied stimulus information during a time period, such during a FI or DRL (differential reinforcement for low rate of responding) schedule (see, however, Macar, 1980). A start signal occurs at the beginning of a time period, but that is the only external information presented. An easy way to test the attentional-gate model would be occasionally to present varied stimulus information (novel events that have not been learned as relating to the reinforcement schedule) during the time period. Using the peak procedure, we predict a peak-shift right. That is, we expect the peak rate of responding to be temporally displaced, occurring later than when no such information is presented. A colleague has suggested that the opposite - a peak shift left - may instead occur. The reason is that the animal may realize it has been distracted and may respond relatively early so as not to postpone the reinforcer. For two reasons, we reject this prediction. First, we hesitate to attribute complex metacognitive processes to animals. Second, human evidence shows that if a concurrent task demands attention, prospective duration judgment of a primary task is shortened, not lengthened (Brown et al., 1992; Brown & West, 1990; Macar et al., 1994).

Why have nearly all animal researchers failed to recognize the importance of attention to time? In the case of those proposing internal-clock models, the answer seems to be that they have failed to distinguish between general arousal and attentional resources (Kahneman, 1973). Proposing a separate role for attention in a modified version of a scalar-timing model has the advantage of parsimoniously accounting for some past findings. For example, Wilkie (1987) varied intensity of a 2-s or 10-s light cue in a choice response task. For equal durations of the dim light and the bright light, pigeons were more likely to choose the *short* response alternative following the dim light, as if the perceived duration of a dim light is shorter than that of a bright light of equal duration. Wilkie suggested that stimulus intensity affects the rate of the pacemaker, an explanation in terms of general arousal. The problem is that this explanation introduces the possibility that a wide variety of variables affect a process which most scalar-timing theorists assume is relatively autonomous, the process underlying the rate of the pacemaker. It is unlikely that the differences in intensity which Wilkie used would lead to different states of arousal, although Treisman (1993) also seemed to assume such an effect. We propose another explanation: intensity of the light influenced the pigeons' allocation of attentional resources.

Contextual-change model

The processes underlying prospective duration judgment differ from those underlying retrospective duration judgment (e.g., Block, 1992; Hicks, Miller, & Kinsbourne, 1976; Zakay, 1990; 1993).

Prospective duration judgment

In the prospective paradigm, subjects are aware that they are engaged in a time-estimation task. All of the animal research and most of the human research on duration timing have used this paradigm.[2] In addition to duration length itself, the most important factor influencing prospective duration judgments is the amount of attention to time that the subject allocates during the duration. If, for example, a concurrent information-processing task is relatively easy, a subject can allocate more attentional resources to time, as opposed to stimulus information (see, for example, Block, 1992; Brown, 1985; Brown & Stubbs, 1992; Brown & West, 1990; Macar et al., 1994; Thomas & Cantor, 1975; Thomas & Weaver, 1975).

Block (1992) recently proposed a contextual-change hypothesis of prospective duration judgment. The most important kind of information influencing duration judgments is varied contextual associations, which may serve as time-tags. In the prospective paradigm, whenever a subject allocates attention to time, contextual information concerning the previous act of attending to time is automatically retrieved, and a new time-tag (set of contextual associations) is encoded. Prospective duration judgment involves estimating the availability of the changes in these time-tags, or temporal context changes.

With only a slight modification, the contextual-change model can be seen as functionally isomorphic with the attentional-gate model developed in the previous section. Figure 7 (p. 186) shows the relabeling of components illustrating this isomorphism. The pacemaker becomes the *context generator*, the cognitive timer becomes the *context recorder*, and the cognitive comparison process becomes the *context comparison* process. The labels for other components remain the same. One major difference is that the contextual information produced by the context generator is not equivalent from moment to moment, at least not in the way that each pulse is assumed to be identical to every other pulse. The

---

[2] However, see Zimmer, this volume, chapter 3.

context comparison process may rely on the total number of unique contextual associations that were encoded during the duration and that are available to a memory retrieval process (cf. Block, 1992). The main advantage of this model over the attentional-gate model is that it reveals more explicit connections with other temporal judgment tasks. For example, judgments of the recency or serial position of an event, as well as the spacing of repeated or related events, seem to depend on contextual associations (Hintzman et al., 1973; Hintzman et al., 1975).

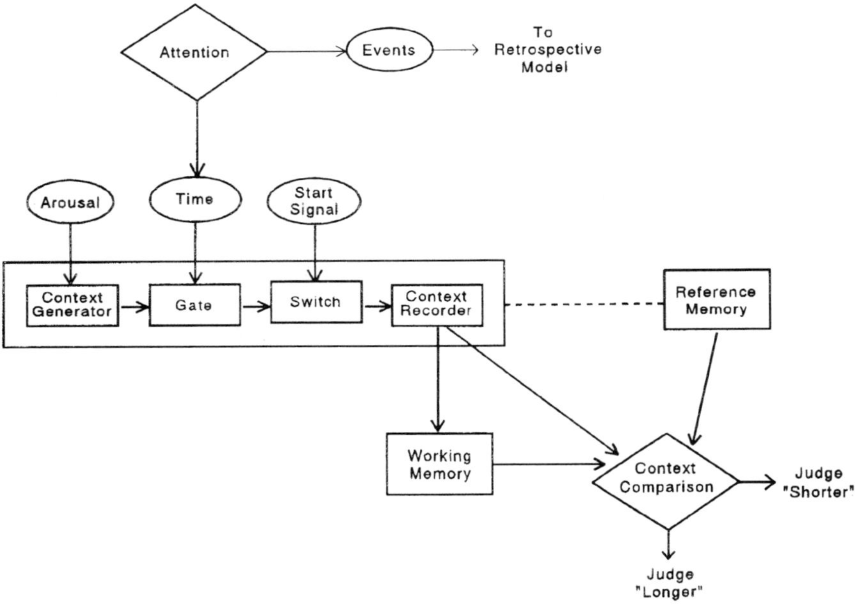

Figure 7: The contextual-change hypothesis of prospective duration judgment diagrammed as a model (after Block, 1992).

Retrospective duration judgment

In contrast to the paradigms we have already discussed, in a retrospective duration-judgment paradigm the person is not aware until after a duration has ended that the situation requires a duration judgment. This is a difficult, if not impossible, paradigm to use in nonhuman animal studies (Wearden & Lejeune, 1993). We cannot easily give animals a short verbal instruction, following a duration, to make a retrospective judgment of the duration. The required training would introduce a lengthy delay between the duration and the animal's judgment. We can ask humans, however, to make such a judgment. Attention to time has little or no influence on retrospective duration judgment. Retrospective judgments do not depend so much on retrieval of temporal context information as on retrieval of other kinds contextual information. This contextual information is encoded in association with event information. It includes environmental, emotional, process, and other similar information. Block (1982; 1985; 1990; 1992; Block & Reed, 1978) proposed a contextual-change hypothesis of retrospective duration judgment, or remembered duration. The remembered duration of a time period lengthens as a function of the amount of contextual changes stored in memory and available to be retrieved at the time of the duration judgment.

Figure 8 (p. 187) shows the components of the attentional-gate model (Figure 6, p. 182) or the contextual-change model of prospective duration judgment (Figure 7, p. 186) needed in a model of retrospective duration judgment. The main focus is on associations formed, mostly automatically, as a subject attends to events (internal or external). The context generator supplies contextual information, which is associated with *event* information and stored in *long-term episodic memory*.

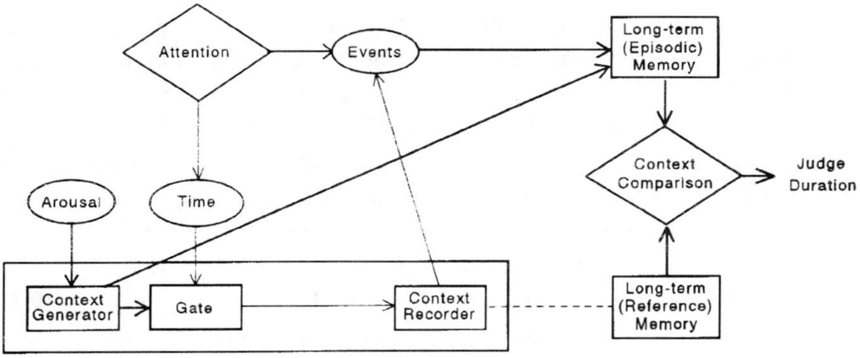

Figure 8: The contextual-change hypothesis of retrospective duration judgment diagrammed as a model (after Block, 1992; Block & Reed, 1978).

Even in a retrospective paradigm, the subject occasionally attends to time. On these relatively rare occasions, the *context recorder* holds information about contextual changes and supplies this information, again in the form of an association to concurrent events. Information about contextual changes is also sent to a *long-term reference memory*. This component holds information about the average amount of unique contextual information stored during durations of various length. In other words, it contains information about the translation from contextual information into duration judgments (expressed verbally or otherwise). Retrospective duration judgments involve a *context comparison* involving this information in long-term episodic memory and in long-term reference memory.

Biopsychological evidence

A variety of biopsychological evidence from both animal and human experiments relates to the kinds of models discussed and proposed here (Block, 1995; Church, 1989). This evidence allows a tentative separation of brain modules or areas subserving the timer from those subserving memory, as well as attentional processes.

Functioning of the internal clock or cognitive timer seems to rely mainly on the frontal lobes of the cerebrum, especially the dorsolateral prefrontal cortex. Converging evidence from psychopharmacological manipulations, electrophysiological recordings, and neuropsychological observations seemingly isolates the timer to this brain region. Researchers who have administered various drugs to animals trained on FI (i.e., peak procedure) and DRL schedules suggest that dopaminergic neurons, which the prefrontal cortex is known to contain, subserve the timer. For example, administering methamphetamine leads to a peak-shift left, as if the animal expected reinforcement sooner. The typical interpretation is that the rate of a neural pacemaker has increased, thereby leading to a greater accumulation of pulses in working memory. Administering dopamine antagonists, such as haloperidol (which blocks postsynaptic dopamine receptors), conversely leads to a peak-shift right. Single-cell recording from neurons in the prefrontal cortex reveals some that are active in the interval between the onset of a stimulus and the time an organism may emit a response (for reviews, see Fuster, 1987; 1989). In humans, damage to this region of the prefrontal cortex disrupts various timing functions, including discriminating the recency and temporal order of events (Milner et al., 1985; 1990; 1991). However, none of this evidence shows that it is necessary to postulate an autonomously functioning, repetitive pacemaker as the first component in duration timing, such as in several models we have discussed here (Figure 3, p. 176, Figure 4, p. 177, and Figure 6, p. 182). As Milner et al. (1990) argue, the evidence may be more consistent with the notion that the dorsolateral prefrontal cortex automatically generates contextual information which may serve as time tags (Figure 7, p. 186, and Figure 8, p. 187).

Other evidence reveals that other brain regions mediate reference memory for temporal (duration) information. The hippocampus and associated medial temporal lobe structures are influenced by cholinergic agonists and antagonists. Administering drugs that influence cholinergic neurons shortens or lengthens the remembered duration of a time period.

For example, anticholinesterases (e.g., physostigmine) and cholinergic receptor blockers (e.g., atropine) influence rats' performance in the peak procedure in ways that scalar timing theory can elegantly handle (Church, 1989; Meck & Church, 1987): The pacemaker rate (Λ) does not change, but the memory storage constant (K*) does. The parameter K* is a bias on the transfer of pulses from the accumulator to reference memory, and the value of K* may be greater or less than 1 depending on such influences as cholinergic drugs. In humans, the remembered duration of a time period is shortened or lengthened in similar ways (Hicks, 1992). Studies of individuals with damage to the medial temporal lobe, especially the hippocampus, provide converging evidence that this brain region is intimately involved in reference memory functions (Block, 1995).

Exactly which areas of the brain subserve attention to time remains unclear. Studies using positron emission tomography reveal that several anatomically separate areas of the human brain, including the thalamus, the parietal lobes, and the anterior cingulate gyrus, play various roles in the performance of attention-demanding tasks (for a review, see Posner & Raichle, 1994). These areas subserve somewhat different functions, which are just beginning to be clarified. The likely candidate for an area subserving the allocation of attention to external events or to time (as in the models shown in Figure 5, p. 181, Figure 6, p. 182, Figure 7, p. 186, and Figure 8, p. 187) is the anterior cingulate gyrus. This area seems to be the central component of an executive attention network, which may directly influence the working-memory functions of the dorsolateral prefrontal cortex. However, the present evidence is too incomplete to suggest anything more definitive about brain components subserving the role of attention in the cognitive models of time we have reviewed and proposed here.

Summary and conclusions

Timing-with-a-timer models assert that a pacemaker, part of an internal clock, underlies psychological timing. Timing-without-a-timer models propose instead that psychological time is constructed from processed and stored information. The scalar-timing model, the best example of timing with a timer, can explain much of the animal data and some of the human data on time-related behavior and judgment. However, it is not easily able to explain why cognitive factors influence temporal behavior and judgment. In order to handle these factors, we proposed a modified model that incorporates an attentional process. This model, the attentional-gate model, is needed to explain findings in which humans divide attention between temporal and nontemporal information. It also explains and predicts some analogous findings in animal research.

The attentional-gate model is roughly isomorphic with a contextual-change model of prospective duration judgment. This model, which involves timing without a timer, replaces the pacemaker mechanism with a process that generates varied contextual information. Temporal context changes, stored as contextual associations with ongoing events, may therefore underlie prospective duration judgments. A modification of this model can also explain retrospective duration judgments, which are more typically explained by proposing cognitive models.

Biopsychological evidence from both animal and human experiments relates to the models reviewed. The areas of the brain that are heavily implicated in various aspects of time-related behavior and judgment include the dorsolateral prefrontal cortex, the anterior cingulate gyrus, and the hippocampus. At present, biopsychological evidence does not unequivocally lead to acceptance or outright rejection of any of the models reviewed here. Future research, using or combining behavioral, cognitive, and biopsychological methods, may clarify the processes underlying time-related behavior and judgment in animals and humans.

References

Allan, L.G. (1992). The internal clock revisited. In F. Macar, V. Pouthas, & W.J. Friedman (Eds.), *Time, action and cognition: towards bridging the gap* (pp. 191-202). Dordrecht, Netherlands: Kluwer Academic.

Block, R.A. (1982). Temporal judgments and contextual change. *Journal of Experimental Psychology: Learning, Memory, and Cognition, 8*, 530-544.

Block, R.A. (1985). Contextual coding in memory: Studies of remembered duration. In J.A. Michon & J.L. Jackson (Eds.), *Time, mind, and behavior* (pp. 169-178). Berlin: Springer.

Block, R.A. (1990). Models of psychological time. In R.A. Block (Ed.), *Cognitive models of psychological time* (pp. 1-35). Hillsdale, NJ: Erlbaum.

Block, R.A. (1992). Prospective and retrospective duration judgment: the role of information processing and memory. In F. Macar, V. Pouthas, & W.J. Friedman (Eds.), *Time, action and cognition: towards bridging the gap* (pp. 141-152). Dordrecht, Netherlands: Kluwer Academic.

Block, R.A. (1995). Psychological time and memory systems of the brain. In J.T. Fraser & M.P. Soulsby (Eds.), *Dimensions of time and life: the study of time VIII* (pp. 61-76). Madison, CT: International Universities Press.

Block, R.A., & Reed, M.A. (1978). Remembered duration: Evidence for a contextual-change hypothesis. *Journal of Experimental Psychology: Human Learning & Memory, 4*, 656-665.

Brown, S.W. (1985). Time perception and attention: the effects of prospective versus retrospective paradigms and task demands on perceived duration. *Perception & Psychophysics, 38*, 115-124.

Brown, S.W., & Stubbs, D.A. (1992). Attention and interference in prospective and retrospective timing. *Perception, 21*, 545-557.

Brown, S. W., Stubbs, D. A., & West, A. N. (1992). Attention, multiple timing, and psychophysical scaling of temporal judgments. In F. Macar, V. Pouthas, & W.J. Friedman (Eds.), *Time, action and cognition: towards bridging the gap* (pp. 129-140). Dordrecht, Netherlands: Kluwer Academic.

Brown, S.W., & West, A.N. (1990). Multiple timing and the allocation of attention. *Acta Psychologica, 75*, 103-121.

Campbell, S.S., & Birnbaum, J.M. (1994). Time flies when you're cool: relationship between body temperature and estimated interval duration [Abstract]. *Sleep Research, 23*, 483.

Church, R.M. (1984). Properties of the internal clock. *Annals of the New York Academy of Sciences, 423*, 566-582.

Church, R.M. (1989). Theories of timing behavior. In S.B. Klein & R.R. Mowrer (Eds.), *Contemporary learning theories: instrumental conditioning theory and the impact of biological constraints on learning* (pp. 41-71). Hillsdale, NJ: Erlbaum.

Doob, L.W. (1971). *Patterning of time*. New Haven, CT: Yale University Press.

Fraisse, P. (1963). *The psychology of time* (transl. by J. Leith). New York: Harper & Row. (Original work published 1957 as *Psychologie du temps*)

François, M. (1927). Contribution à l'étude du sens du temps: La température interne comme facteur de variation de l'appréciation subjective des durées [Contribution to the study of the time sense: Internal temperature as a factor varying the subjective appreciation of duration]. *Année Psychologique, 27*, 186-204.

Frankenhaeuser, M. (1959). *Estimation of time: an experimental study*. Stockholm: Almqvist & Wiksell.

Fuster, J.M. (1987). Single-unit studies of the prefrontal cortex. In E. Perecman (Ed.), *The frontal lobes revisited* (pp. 109-120). New York: IRBN Press.

Fuster, J.M. (1989). *The prefrontal cortex: anatomy, physiology and neuropsychology*. New York: Raven Press.

Gibbon, J. (1977). Scalar expectancy theory and Weber's law in animal timing. *Psychological Review, 84*, 279-325.

Gibbon, J. & Church, R.M. (1984). Sources of variance in an information processing theory of timing. In H.L. Roitblat, T.G. Bever, & H.S. Terrace (Eds.), *Animal cognition* (pp. 465-488). Hillsdale, NJ: Erlbaum.

Gibbon, J., Church, R.M., & Meck, W.H. (1984). Scalar timing in memory. *Annals of the New York Academy of Sciences*, 423, 52-77.

Guyau, J.-M. (1988). La genèse de l'idée de temps [The origin of the idea of time]. In J.A. Michon, V. Pouthas, & J.L. Jackson (Eds.), *Guyau and the idea of time* (pp. 37-90; trans. pp. 93-148). Amsterdam: North-Holland. (Original work published 1890)

Hancock, P.A. (1993). Body temperature influence on time perception. *Journal of General Psychology, 120*, 197-216.

Hicks, R.E. (1992). Prospective and retrospective judgments of time: A neurobehavioral analysis. In F. Macar, V. Pouthas, & W.J. Friedman (Eds.), *Time, action and cognition: towards bridging the gap* (pp. 97-108). Dordrecht, Netherlands: Kluwer Academic.

Hicks, R.E., Miller, G.W., & Kinsbourne, M. (1976). Prospective and retrospective judgments of time as a function of amount of information processed. *American Journal of Psychology, 89*, 719-730.

Hintzman, D.L., Block, R.A., & Summers, J.J. (1973). Contextual associations and memory for serial position. *Journal of Experimental Psychology, 97*, 220-229.

Hintzman, D.L., Summers, J.J., & Block, R.A. (1975). Spacing judgments as an index of study-phase retrieval. *Journal of Experimental Psychology: Human Learning & Memory, 104*, 31-40.

Hoagland, H. (1933). The physiological control of judgments of duration: Evidence for a chemical clock. *Journal of General Psychology*, *9*, 267-287.

Hoagland, H. (1966). Some biochemical considerations of time. In J.T. Fraser (Ed.), *The voices of time* (pp. 312-329). New York: Braziller.

Ivry, R.B., & Hazeltine, R.E. (1992). Models of timing-with-a-timer. In F. Macar, V. Pouthas, & W.J. Friedman (Eds.), *Time, action and cognition: towards bridging the gap* (pp. 183-189). Dordrecht, Netherlands: Kluwer Academic.

Kahneman, D. (1973). *Attention and effort*. Englewood Cliffs, NJ: Prentice-Hall.

Lashley, K.S. (1951). The problem of serial order in behavior. In L.A. Jeffress (Ed.), *Cerebral mechanisms in behavior: the Hixon symposium* (pp. 112-146). New York: Wiley.

Macar, F. (1980). Physiological mechanisms. In M. Richelle & H. Lejeune (Eds.), *Time in animal behaviour* (pp. 143-165). Oxford, England: Pergamon Press.

Macar, F., Grondin, S., & Casini, L. (1994). Controlled attention sharing influences time estimation. *Memory & Cognition*, *22*, 673-686.

Meck, W.H. (1984). Attentional bias between modalities: effect on the internal clock, memory, and decision stages used in animal time discrimination. *Annals of the New York Academy of Sciences*, *423*, 528-541.

Meck, W.H., & Church, R.M. (1987). Cholinergic modulation of the content of temporal memory. *Behavioral Neuroscience*, *101*, 457-464.

Michon, J.A. (1985). The complete time experiencer. In J.A. Michon & J.L. Jackson (Eds.), *Time, mind, and behavior* (pp. 20-52). Berlin: Springer.

Milner, B., Corsi, P., & Leonard, G. (1991). Frontal-lobe contribution to recency judgements. *Neuropsychologia*, *29*, 601-618.

Milner, B., McAndrews, M.P., & Leonard, G. (1990). Frontal lobes and memory for the temporal order of recent events. *Cold Spring Harbor Symposia on Quantitative Biology*, *55*, 987-994

Milner, B., Petrides, M., & Smith, M.L. (1985). Frontal lobes and the temporal organization of memory. *Human Neurobiology*, *4*, 137-142.

Ornstein, R.E. (1969). *On the experience of time*. Harmondsworth, England: Penguin.

Piéron, H. (1923). Les problèmes psychophysiologiques de la perception du temps. *Année Psychologique*, *24*, 1-25.

Posner, M.I., & Raichle, M.E. (1994). *Images of mind*. New York: Scientific American Library.

Pouthas, V. (1985). Timing behavior in young children: A developmental approach to conditioned spaced responding. In J.A. Michon & J.L. Jackson (Eds.), *Time, mind, and behavior* (pp. 100-109). Berlin: Springer.

Richelle, M., & Lejeune, H. (1980). Time in animal behavior. Oxford, England: Pergamon Press.

Richelle, M., Lejeune, H., Perikel, J.-J., & Fery, P. (1985). From biotemporality to nootemporality: toward an integrative and comparative view of time in behavior. In J.A. Michon & J.L. Jackson (Eds.), *Time, mind, and behavior* (pp. 75-99). Berlin: Springer.

Roberts, S., & Church, R.M. (1978). Control of an internal clock. *Journal of Experimental Psychology: Animal Behavior Processes*, *4*, 318-337.

Roberts, S. (1983). Properties and function of an internal clock. In R.L. Mellgren (Ed.), *Animal cognition and behavior* (pp. 345-397). Amsterdam: North-Holland.

Thomas, E.A.C., & Cantor, N.E. (1975). On the duality of simultaneous time and size perception. *Perception & Psychophysics*, *18*, 44-48.

Thomas, E.A.C., & Weaver, W.B. (1975). Cognitive processing and time perception. *Perception & Psychophysics*, *17*, 363-367.

Treisman, M. (1963). Temporal discrimination and the indifference interval: Implications for a model of the „internal clock". *Psychological Monographs*, *77*, (13, Whole No. 576), 1-13.

Treisman, M. (1984). Temporal rhythms and cerebral rhythms. *Annals of the New York Academy of Sciences*, 423, 542-565.

Treisman, M. (1993). On the structure of the temporal sensory system. *Psychologica Belgica*, 33, 271-283.

Treisman, M., Faulkner, A., Naish, P.L., & Brogan, D. (1990). The internal clock: evidence for a temporal oscillator underlying time perception with some estimates of its characteristic frequency. *Perception, 19*, 705-743.

Wearden, J.H. & Lejeune, H. (1993). Across the great divide: animal psychology and time in humans. *Time and Society, 2*, 87-106.

Wearden, J.H., & Penton-Voak, I.S. (1995). Feeling the heat: body temperature and the rate of subjective time, revisited. *Quarterly Journal of Experimental Psychology: Comparative and Physiological Psychology, 48*, 129-141.

Wilkie, D. M. (1987). Stimulus intensity affects pigeons' timing behavior: implications for an internal clock model. *Animal Learning and Behavior*, 15, 35-39.

Zakay, D. (1990). The evasive art of subjective time measurement: some methodological dilemmas. In R.A. Block (Ed.), *Cognitive models of psychological time* (pp. 59-84). Hillsdale, NJ: Erlbaum.

Zakay, D. (1993). Relative and absolute duration judgments under prospective and retrospective paradigms. *Perception & Psychophysics, 1*, 656-664.

Zakay, D., & Block, R. (1994, November). A functional model of the cognitive timer. Paper presented at the meeting on Time and the Dynamic Control of Behavior, Liège, Belgium.

Subject Index

A

B

C

D

J

K

L

M

N

O

P

U

V

W

Z

Author Index

K

L

M

N

O

P

R

S

T

U

V

W

Y

Z